T0224617

Lecture Notes in Mathematics

Edited by J.-M. Morel, F. Takens and B. Teissier

Editorial Policy
for the publication of monographs

1. Lecture Notes aim to report new developments in all areas of mathematics – quickly, informally and at a high level. Monograph manuscripts should be reasonably self-contained and rounded off. Thus they may, and often will, present not only results of the author but also related work by other people. They may be based on specialized lecture courses. Furthermore, the manuscripts should provide sufficient motivation, examples and applications. This clearly distinguishes Lecture Notes from journal articles or technical reports which normally are very concise. Articles intended for a journal but too long to be accepted by most journals, usually do not have this "lecture notes" character. For similar reasons it is unusual for doctoral theses to be accepted for the Lecture Notes series.

2. Manuscripts should be submitted (preferably in duplicate) either to one of the series editors or to Springer-Verlag, Heidelberg. In general, manuscripts will be sent out to 2 external referees for evaluation. If a decision cannot yet be reached on the basis of the first 2 reports, further referees may be contacted: the author will be informed of this. A final decision to publish can be made only on the basis of the complete manuscript, however a refereeing process leading to a preliminary decision can be based on a pre-final or incomplete manuscript. The strict minimum amount of material that will be considered should include a detailed outline describing the planned contents of each chapter, a bibliography and several sample chapters.
Authors should be aware that incomplete or insufficiently close to final manuscripts almost always result in longer refereeing times and nevertheless unclear referees' recommendations, making further refereeing of a final draft necessary.
Authors should also be aware that parallel submission of their manuscript to another publisher while under consideration for LNM will in general lead to immediate rejection.

3. Manuscripts should in general be submitted in English.
Final manuscripts should contain at least 100 pages of mathematical text and should include
– a table of contents;
– an informative introduction, with adequate motivation and perhaps some historical remarks: it should be accessible to a reader not intimately familiar with the topic treated;
– a subject index: as a rule this is genuinely helpful for the reader.

Continued on inside back-cover

Lecture Notes in Mathematics   1792

Editors:
J.-M. Morel, Cachan
F. Takens, Groningen
B. Teissier, Paris

Springer
*Berlin
Heidelberg
New York
Barcelona
Hong Kong
London
Milan
Paris
Tokyo*

Dang Dinh Ang
Rudolf Gorenflo
Vy Khoi Le
Dang Duc Trong

Moment Theory and Some Inverse Problems in Potential Theory and Heat Conduction

Springer

Authors

Dang Dinh ANG
Department of Mathematics
and Informatics
HoChiMinh City National University
227 Nguyen Van Cu, Q5
Ho Chi Minh City
Viet Nam
e-mail: khanhchu@mail.saigonnet.vn

Rudolf GORENFLO
Department of Mathematics
and Informatics
Free University of Berlin
Arnimallee 3
14195 Berlin
Germany
e-mail: gorenflo@math.fu-berlin.de
http://www.fracalmo.org

Vy Khoi LE
Department of Mathematics
and Statistics
University of Missouri-Rolla
Rolla, Missouri 65401
USA
e-mail: vy@umr.edu

Dang Duc TRONG
Department of Mathematics
and Informatics
HoChiMinh City National University
227 Nguyen Van Cu, Q5
Ho Chi Minh City
Viet Nam
e-mail:
ddtrong@mathdep.hcmuns.edu.vn

Cataloging-in-Publication Data applied for.

Die Deutsche Bibliothek - CIP-Einheitsaufnahme

Moment theory and some inverse problems in potential theory and heat
conduction / Dang Dinh Ang .... - Berlin ; Heidelberg ; New York ; Barcelona
; Hong Kong ; London ; Milan ; Paris ; Tokyo : Springer, 2002
(Lecture notes in mathematics ; 1792)
ISBN 3-540-44006-2

Mathematics Subject Classification (2000):
30E05, 30E10, 31A35, 31B20, 35R25, 35R30, 44A60, 45Q05, 47A52

ISSN 0075-8434
ISBN 3-540-44006-2 Springer-Verlag Berlin Heidelberg New York

Springer-Verlag Berlin Heidelberg New York a member of BertelsmannSpringer
Science + Business Media GmbH

http://www.springer.de

© Springer-Verlag Berlin Heidelberg 2002
Printed in Germany

Typesetting: Camera-ready TEX output by the author

SPIN: 10884684   41/3142/ du - 543210 - Printed on acid-free paper

Foreword

In recent decades, the theory of inverse and ill-posed problems has impressively developed into a highly respectable branch of Applied Mathematics and has had stimulating effects on Numerical Analysis, Functional Analysis, Complexity Theory, and other fields. The basic problem is to draw useful information from noise contaminated physical measurements, where in the case of ill-posedness, naive methods of evaluation lead to intolerable amplification of the noise. Usually, one is looking for a function (defined on a suitable domain) that is close to the true function assumed to exist as underlying the situation or process the measurements are taken from, and the above mentioned gross amplification of noise (mathematically often caused by the attempt to invert an operator whose inverse is unbounded) makes the numerical results so obtained useless, these "results" hiding the true solution under large amplitude high frequency oscillations.

There is an ever growing literature on ways out of this dilemma. The way out is to suppress unwanted noise, thereby avoiding excessive suppression of relevant information. Various methods of "regularization" have been developed for this purpose, all, in principle, using extra information on the unknown function. This can be in the form of general assumptions on "smoothness", an idea underlying, e.g., the method developed by Tikhonov and Phillips (minimization of a quadratic functional containing higher derivatives in an attempt to reproduce the measured data) and various modifications of this method. Another efficient method is the so-called "regularization by discretization" method where one has to find a kind of balance between the fineness of discretization and its tendency to amplify noise. Yet another method, the so-called "descriptive regularization" method, consists in exploiting a priori known characteristics of the unknown function, such as regions of nonnegativity, or monotonicity, or convexity that can be used in a scheme of linear or nonlinear fitting to the measured data, fitting optimal with respect to appropriate constraints. Many ramifications and combinations of these and other methods have been analyzed theoretically and used in numerical calculations. Our monograph deals with the method called the "moment method". The moments considered here are of the form

$$\mu_n = \int_\Omega u(x)d\sigma_n, \qquad n = 1, 2, 3, ...,$$

where Ω is a domain in $\mathbf{R}^k$, $d\sigma_n$ is, either a Dirac measure, $n \in \mathbf{N}$, or a measure absolutely continuous with respect to the Lebesgue measure, i.e.,

$$d\sigma_n = g_n(x)dx, \qquad n \in \mathbf{N},$$

$g_n(x)$ being Lebesgue integrable on Ω. The idea of the moment method is to reconstruct an unknown function $u(x)$ from a given set $(\mu_n)_{n \in I}$, $I \subset \mathbf{N}$, of the moments of $u(x)$. Then the problem arises as to whether a knowledge of moments of $u(x)$ uniquely determines this function. For the moment problems considered in this monograph, unless stated otherwise, the knowledge of the complete sequence of moments of $u(x)$ uniquely determines the function. In practice, one has available only a finite set $\mu_1, ..., \mu_m$ of moments, and furthermore these are usually contaminated with noise, the reason being that they are results of experimental measurements. The question then is: To what extent, can the true function $u(x)$ be recovered from the finite set $(\mu_i)_{1 \leq i \leq m}$ of moments? Note that in the latter situation, the question of existence of a solution u plays a minor role. The moments being only approximately known, the problem is reduced to one of "regularization", namely, to the problem of fitting the function $u(x)$ as closely as possible to the available data, that is, to the given approximate values of the moments, $u(x)$ being assumed to lie in a nice function space and to obey a known or stipulated restriction to the size of an appropriate functional. In our theory of regularization, the index m, i.e., the number of the given moment values mentioned above, will play the role of the regularization parameter. In illustration of the theory, we shall study several concrete cases, discussing inverse problems of function theory, potential theory, heat conduction and gravimetry. We will make essential use of analyticity or harmonicity of the functions involved, and so the theory of analytic functions and harmonic functions will play a decisive role in our investigations. We hope that this monograph, which is a fruit of several years of joint efforts, will stimulate further research in theoretical as well as in practical applications.

It is our pleasure to acknowledge with gratitude the valuable assistance of several researchers with whom we could discuss aspects of the theory of moments, either after presentation in conferences and seminars or in personal exchange of knowledge and opinions. Special thanks are due to our colleagues Johann Baumeister, Bernd Hofmann, Sergio Vessella, Lothar von Wolfersdorf and Masahiro Yamamoto. They have studied the whole manuscript and their detailed constructive-critical remarks have helped us much in improving it. Our thanks are also due to the anonymous referees for their valuable suggestions. Last not least, we highly appreciate the supports granted by Deutsche Forschungsgemeinschaft in Bonn which made possible several mutual research visits, furthermore the supports given by the Research Commission of Free University of Berlin, Ho Chi Minh City Mathematical Society, Ho Chi Minh City National University, and the Vietnam Program of Basic Research in the Natural Sciences. Last not least we are grateful to Ms. Julia Loutchko for her help in the final corrections and preparations of the manuscript for publishing.

Dang Dinh Ang, Rudolf Gorenflo,
Vy Khoi Le and Dang Duc Trong

Berlin, Ho Chi Minh City, Rolla-Missouri: March 2002

Table of Contents

Introduction

A moment problem is either a problem of finding a function u on a domain Ω of $\mathbf{R}^d$, $d \geq 1$, satisfying a sequence of equations of the form

$$\int_\Omega u\, d\sigma_n = \mu_n \tag{0.1}$$

where $(d\sigma_n)$ is a given sequence of measures on Ω and (μ_n) is a given sequence of numbers, or a problem of finding a measure $d\sigma$ on Ω satisfying a sequence of equations of the form

$$\int_\Omega g_n\, d\sigma = \mu_n, \tag{0.2}$$

for given g_n and μ_n, $n = 1, 2, \dots$. Although this monograph is devoted exclusively to a study of moment problems of the form (0.1), we shall briefly mention a classical result on moment problems of the form (0.2) in the *Notes and Remarks* of Chapter 2. Concerning moment problems of the form (0.1), if $d\sigma_n$ is absolutely continuous with respect to the Lebesgue measure, i.e., if

$$d\sigma_n = g_n dx,$$

where g_n is Lebesgue integrable, n=1,2,..., then we have the usual moment problem

$$\int_\Omega u g_n dx = \mu_n. \tag{0.3}$$

If $d\sigma_n$ is a Dirac measure, i.e., if

$$d\sigma_n = \delta(x - x_n), \qquad x_n \in \Omega, \tag{0.4}$$

then the moment problem consists in finding a function u on Ω from its values at a sequence of points (x_n), i.e.,

$$u(x_n) = \mu_n, \qquad n = 1, 2, \dots . \tag{0.5}$$

Before proceeding further, it seems appropriate to explain how each of the two foregoing variants of the moment problem (0.1) arises in the framework of this monograph. In fact, many inverse problems can be formulated as an integral equation of the first kind, namely,

$$\int_a^b K(x, y) u(y) dy = f(x), \qquad x \in (a, b), \tag{0.6}$$

where (a, b) is a bounded or unbounded open interval of $\mathbf{R}$. Here $K(x, y)$ and $f(x)$ are given functions and $u(y)$ is a solution to be determined. In practice, $f(x)$ is a result of experimental measurements and hence is given only at a finite set of points that is conveniently patched up into a continuous function or an L^2-function. This is an interpolation problem. Interpolation is a delicate process, and, in general, it is difficult to know the number of points needed to achieve a desired degree of approximation unless the function $f(x)$ is sufficiently smooth. The case that the function represented by the integral in the above equation can be extended to a function complex analytic in a strip of the complex plane $\mathbf{C}$ containing the real interval $[a, b]$ is of special interest. Indeed, under the analyticity assumption, if the left hand side of the equation is known on a bounded sequence (x_n) in (a, b) with $x_i \neq x_j$ for $i \neq j$, then by a well-known property of analytic functions, the function is known in the strip and a fortiori in (a, b). It follows that the above integral equation is equivalent to the following moment problem

$$\int_a^b K(x_n, y)u(y)dy = f(x_n), \qquad n = 1, 2, \dots . \tag{0.7}$$

In some examples to be given in later chapters, we also have moment problems of the foregoing form with (x_n) unbounded and satisfying certain properties. We shall also deal with multidimensional moment problems

$$\int_\Omega K(x_n, y)u(y)dy = f(x_n), \qquad n = 1, 2, \dots \tag{0.8}$$

where Ω is a domain in $\mathbf{R}^d$, $d \geq 1$ and (x_n) is some infinite sequence (not necessarily in Ω).

As mentioned earlier, we can have moment problems of the form (0.1) above, with the $d\sigma_n$'s being Dirac measures. This moment problem will arise in the reconstruction of a function u analytic in the unit disc U of $\mathbf{C}$ from its values at a given sequence of points (z_n) of U,

$$u(z_n) = \mu_n, \qquad n = 1, 2, \dots . \tag{0.9}$$

Moment problems are similar to integral equations except that we now deal with mappings between different spaces. Hence special techniques are required.

The purpose of this monograph is to present some basic techniques for treatments of moment problems. We note that classical treatments are concerned primarily with questions of existence (and uniqueness). For the classical theory, the reader is referred to, e.g., the monograph of Akhiezer [Ak] and the article of Landau [La]. From our point of view, however, the given data are results of experimental measurements and hence are given only at finite sets of points that are conveniently patched up into functions in appropriate spaces, and consequently, a solution may not exist. Furthermore, moment problems are ill-posed in the sense that solutions usually do not exist and that in the case of existence, there is no continuous dependence on the given data. The present monograph presents some regularization methods.

Parallel to the theory of moments, we shall consider various inverse problems in Potential Theory and in Heat Conduction. These inverse problems provide important examples in illustration of moment theory, however, they are also investigated for their own sake. In order to convey the full flavor of the subject, we have tried to explain in detail the physical models.

The book consists of seven chapters. The first five chapters deal with mathematical preliminaries (Chapter 1) and mathematical aspects of moment theory

(Chapters 2 to 5). The remaining two chapters are devoted to concrete inverse problems in Potential Theory and in Heat Conduction.

Chapter 1 contains the mathematical preliminaries in preparation for the subsequent chapters. Chapter 2 presents various methods of regularization for moment problems: the method of truncated expansion and the method of Tikhonov in Hilbert spaces and in reflexive Banach spaces. Chapter 3 is devoted to the Backus-Gilbert theory in Hilbert spaces and in reflexive Banach spaces. Chapter 4 deals with the Hausdorff moment problem in one dimension and in several dimensions. Chapter 5 deals with the reconstruction of an analytic function in the unit disc using approximations by finite moments (i.e. by a finite set of values of moments) and the method of optimal recovery. In the same chapter, we establish a theorem on cardinal series representation in the two-dimensional case and a theorem of approximation by Sinc functions. The results of Chapter 5 are used repeatedly in subsequent chapters.

The last two chapters of the book deal with some inverse problem in Applied Sciences. Chapter 6 presents some basic properties of harmonic functions and treatments of various regularization methods for Cauchy's problem with applications in Medicine and Geophysics. Chapter 7 is concerned with some inverse problems in heat conduction (the backward heat equation, the problem of surface temperature determination from borehole measurements, the inverse Stefan problem) and presents some methods of regularization for these problems. The book closes with an Epilogue giving an example of a nonlinear moment problem from Gravimetry.

For some chapters, under the heading "Notes and remarks", results are presented as supplements to the main text. At the end of the book, there is a bibliography on all the topics covered in the volume.

This monograph is an introduction to the theory of moments and to some inverse problems in the physical sciences formulated as moment problems. It is not meant to be an exhaustive treatment of moment theory, and we beg pardon, in advance, for the many omissions of important topics (such as, e.g., the maximum entropy method). For further developments in moment theory and in inverse problems in Potential Theory and in Heat Conduction, the reader would do well to consult the references listed at the end of the book as well as the current literature on the subject.

The book can be used as a supplementary text for graduate or advanced undergraduate courses in Inverse Problems or in Mathematical Methods in the Physical Sciences.

1 Mathematical preliminaries

In this short chapter, we collect some results on Banach spaces, in particular on Hilbert spaces, on operator theory, on function spaces (spaces of continuous functions, Lebesgue spaces, Sobolev spaces), on analytic functions, on harmonic functions and on integral transforms (Laplace transform, Fourier transform) for use in subsequent chapters. The results are stated without proof or as consequences of general theorems. References are given to appropriate sources (textbooks or papers).

1.1 Banach spaces

Let X be a Banach space. A subset K of X is called *compact* if each sequence in K has a subsequence converging to an element of K. A subset K is called *relatively compact* if its closure $\overline{K}$ is compact. One has (see, e.g., [Br], page 92)

Theorem 1.1. (Riesz) *Let X be a Banach space such that the open ball $B_1(0)$ centered at 0 with radius 1 is relatively compact. Then X is a finite dimensional vector space.*

Let X, Y be two Banach spaces with respect to the norms $\| \cdot \|_X$, $\| \cdot \|_Y$. We denote by $\mathcal{L}(X, Y)$ the space of all continuous linear operators A from X to Y with the norm

$$\|A\|_{\mathcal{L}(X,Y)} = \sup_{\|x\|_X \leq 1} \|Ax\|_Y$$

With the latter norm, $\mathcal{L}(X, Y)$ is a Banach space. If $X = Y$ then we denote $\mathcal{L}(X, Y)$ by $\mathcal{L}(X)$. An operator A in $\mathcal{L}(X, Y)$ is said to be *compact* if the set $A(K)$ has compact closure in Y for each bounded set K in X.

If X is a Banach space, we write X^* for $\mathcal{L}(X, \mathbf{C})$, i.e., X^* is the set of all continuous linear functionals on X. If $f \in X^*$, we write

$$< f, x > = f(x) \qquad \text{for } x \text{ in } X,$$
$$\|f\|_{X^*} = \sup_{\|x\|_X \leq 1} | < f, x > |.$$

X^* together with the norm $\| \cdot \|_{X^*}$ is a Banach space and X^* is called *the dual* of X. We set $X^{**} = (X^*)^*$.

Let x be in X. If we put

$$T_x f = <f, x> \qquad \text{for } f \text{ in } X^*.$$

then $T_x : X^* \longrightarrow \mathbf{C}$ is a continuous linear functional, i.e., $T_x \in X^{**}$, and $\|T_x\|_{X^{**}} = \|x\|_X$. Letting $j(x) = T_x$, we obtain an isometric linear map

$$j: X \longrightarrow X^{**}.$$

Since j is injective, we can identify X with the subspace $j(X)$ of X^{**}. The Banach space X is called *reflexive* if $j(X) = X^{**}$. In this case j is an isometry from X onto X^{**}, and we write $X = X^{**}$.

A sequence (x_n) in the Banach space X is said to be *weakly convergent* to x in X if, for all f in X^*,

$$<f, x_n> \longrightarrow <f, x> \qquad \text{as } n \to \infty,$$

and we write

$$x_n \rightharpoonup x \qquad as \ n \to \infty.$$

We have (see, e.g., [Br], p. 44)

Theorem 1.2. (Kakutani) *Each bounded sequence in a reflexive Banach space has a weakly convergent subsequence.*

In Chapter 3, we shall give some special results related to Banach spaces and their duals.

1.2 Hilbert spaces

In the remainder of this chapter, the symbol H denotes a complex Hilbert space. We say that $x \in H$ is *orthogonal* to $y \in H$, and write $x \perp y$, if $(x, y)_H = 0$, where $(.,.)_H$ is the inner product of H. For any linear subspace M of H, put

$$M^\perp = \{y \in H : (x, y)_H = 0 \text{ for all } x \in M\}$$

If $M = \{x\}$ then we denote $M^\perp$ by $x^\perp$. We have (cf. [Ru], Chap. 4 or [Br], Chap. 5)

Theorem 1.3 *Let M be a closed subspace of H. Then there exists a unique pair of continuous linear operators*

$$P: H \longrightarrow M, \quad Q: H \longrightarrow M^\perp$$

such that
  a) $x = Px + Qx,$
  b) *If $x \in M$ then $Px = x$, $Qx = 0$ and if $x \in M^\perp$ then $Px = 0$, $Qx = x$,*
  c) $\|x\|_H^2 = \|Px\|_H^2 + \|Qx\|_H^2,$
and we write

$$H = M \oplus M^{\perp}.$$

From the latter theorem, we deduce Riesz's representation for continuous linear functionals.

Theorem 1.4. *If $f \in H^*$ then there exists a uniquely determined element v of H such that*

$$f(x) = (x, v)_H \qquad \text{for all } x \in H.$$

Moreover $\|f\|_{H^*} = \|v\|_H$.

Consider the bilinear form $a(u, v)$ on H, i.e., a function $a : H \times H \to \mathbf{C}$ such that

$$a(x + \lambda z, y) = a(x, y) + \lambda a(z, y)$$
$$a(x, y + \lambda z) = a(x, y) + \bar{\lambda} a(x, z)$$

for all $x, y, z \in H, \ \lambda \in \mathbf{C}$.

Using Theorem 1.4, we get the following theorem (cf. [Br]).

Theorem 1.5. (Lax-Milgram's theorem) *Let $a : H \times H \to \mathbf{C}$ be a bilinear form. Assume that*

a) a is bounded, i.e., there is a $C > 0$ such that

$$|a(x, y)| \leq C\|x\|_H \|y\|_H \qquad \text{for } x, y \in H,$$

b) a is coercive, i.e., there is a $C_0 > 0$ such that

$$a(x, x) \geq C_0 \|x\|_H^2 \qquad \text{for } x \in H.$$

Then, for each f in H^*, there is a unique element u in H such that

$$a(u, v) = f(v) \qquad \text{for } v \in H.$$

The mapping $f \longmapsto u$ is a one-to-one continuous linear map from H^* onto H.

A set of vectors u_α in a Hilbert space H, where α runs through some index set I, is said to be an *orthonormal basis* of H if

(i) $(u_\alpha, u_\beta)_H = 0$ for all $\alpha \neq \beta, \alpha, \beta$ in I,
(ii) $\|u_\alpha\|_H = 1$ for all $\alpha \in I$,
(iii) the set of all finite linear combinations of members of $\{u_\alpha\}$ is dense in H.

In particular, if $I = \mathbf{N}$, then, the space H has a countable orthonormal basis $\{u_n\}$. In the latter case, one has

Theorem 1.6. (Riesz-Fisher) *Let $\{u_n\}$ be a countable orthonormal basis of H. The element x is in H if and only if there exists a complex sequence (c_n) satisfying $\sum_{n=1}^{\infty} |c_n|^2 < \infty$ such that one has the expansion*

$$x = \sum_{n=1}^{\infty} c_n u_n.$$

Moreover, one has $c_n = (x, u_n)_H$ and

$$\|x\|_H^2 = \sum_{n=1}^{\infty} |c_n|^2.$$

Let H_1, H_2 be two Hilbert spaces with respect to the inner products $(.,.)_{H_1}$, $(.,.)_{H_2}$. If $A: H_1 \longrightarrow H_2$ is a continuous linear operator, then the *adjoint* of A is the operator $A^*: H_2 \longrightarrow H_1$ satisfying

$$(Ax, y)_{H_2} = (x, A^*y)_{H_1} \qquad \text{for all } x \in H_1, y \in H_2.$$

If $H_1 = H_2 = H$ and $A = A^*$, then A is called *self-adjoint*. One has the following *spectral theorem* (see, e.g.,[Br], chap. 6).

Theorem 1.7. *Let H be a Hilbert space having a countable orthonormal basis . If $A: H \longrightarrow H$ is an arbitrary self-adjoint compact operator, then there exists an orthonormal basis $\{e_n\}$ and a real sequence (λ_n) tending to zero such that $Ae_n = \lambda_n e_n$.*

A continuous linear operator $A: H \longrightarrow H$ is called *positive* if

$$(Ax, x)_H \geq 0 \qquad \text{for all } x \in H.$$

One has the following result (see, e.g.,[LS]).

Theorem 1.8. *If $A: H \longrightarrow H$ is an arbitrary positive self-adjoint continuous linear operator, then there exists uniquely a positive continuous linear operator $B: H \longrightarrow H$ such that $B^2 = A$.*

In particular, for A in $\mathcal{L}(H_1, H_2)$, the operator $A^*A: H_1 \longrightarrow H_1$ is a positive self-adjoint continuous linear operator. Hence, Theorem 1.8 implies that there is a unique positive continuous linear operator $C: H_1 \longrightarrow H_1$ such that $C^2 = A^*A$.

1.3 Some useful function spaces

1.3.1 Spaces of continuous functions

Let K be a compact subset of $\mathbf{R}^k$. We denote by $C(K)$ the Banach space of continuous functions f from K to $\mathbf{C}$ with the norm

$$\|f\|_{C(K)} = \sup_{x \in K} |f(x)|.$$

Let D be a bounded domain of $\mathbf{R}^k$, $k \geq 1$. For $m = 1, 2, ...$, we consider the space $C^m(\overline{D})$ $(C^m(D))$ of all functions

$$f : \overline{D} \longrightarrow \mathbf{C} \text{ (or } \mathbf{R})$$

$(f : D \longrightarrow \mathbf{C} \text{ (or } \mathbf{R}))$ such that the derivatives

$$D^\alpha f \equiv \frac{\partial^{|\alpha|} f}{\partial x_1^{\alpha_1} ... \partial x_k^{\alpha_k}}$$

are continuous on $\overline{D}$ (D) for $\alpha = (\alpha_1, ..., \alpha_k)$, $|\alpha| = \alpha_1 + ... + \alpha_k$ and $|\alpha| \leq m$.

We denote by $C^\infty(D)$ the space of functions which are infinitely differentiable. The space $C^m(\overline{D})$ is a Banach space with respect to the norm

$$\|f\|_{C^m(\overline{D})} = \sum_{|\alpha| \leq m} \sup_{x \in \overline{D}} |D^\alpha f(x)|.$$

Let G_0 be in $C(\overline{D} \times \overline{D})$. We have (cf. [Mi], §8, Chap. 2)

Theorem 1.9. *Let D be a bounded domain in $\mathbf{R}^k$. For $0 \leq \alpha < k$, the mapping*

$$Tf(x) = \int_D G_0(x, y)|x - y|^{-\alpha} f(y) dy, \qquad f \in C(\overline{D}),$$

is a compact linear operator on $C(\overline{D})$.

T is called a Fredholm integral operator.

The following theorems give some properties of continuous functions on a compact subset. In fact, one has (cf. [Br], Chap. 4, and [HSt]) the following two theorems.

Theorem 1.10. (Ascoli) *Let K be a compact set in $\mathbf{R}^k$ and let $\mathcal{K}$ be a bounded subset of $C(K)$. Suppose that $\mathcal{K}$ is equicontinuous, i.e., for every $\epsilon > 0$, there exists $\delta > 0$ such that*

$$|f(x) - f(y)| < \epsilon \qquad \text{for all } f \text{ in } \mathcal{K}, \ d(x, y) < \delta, \ x, y \in K,$$

where $d(x, y)$ is the distance between x and y in $\mathbf{R}^k$. Then $\mathcal{K}$ is relatively compact in $C(K)$.

Theorem 1.11. (Dini) *Let K be a compact subset of $\mathbf{R}^k$ and let (f_n) be a monotone sequence in $C(K)$ that converges pointwise to a function f in $C(K)$. Then $f_n \to f$ uniformly on K.*

1.3.2 Spaces of integrable functions

Let X be a measure space with a positive measure μ. For $1 \leq p < \infty$, we denote by $L^p(X, \mu)$ the Banach space of complex measurable functions f on X to $\mathbf{C}$ with respect to the norm

$$\|f\|_{L^p(X,\mu)} = \left(\int_X |f|^p d\mu \right)^{1/p}.$$

For a measurable function f, we put

$$\|f\|_{L^\infty(X,\mu)} = \inf\{\alpha \in \mathbf{R} : \mu(\{x \in X : |f(x)| > \alpha\}) = 0\}.$$

We denote by $L^\infty(X,\mu)$ the Banach space of all measurable functions f satisfying $\|f\|_{L^\infty(X,\mu)} < \infty$.

One has (cf [Ru], Chap. 1)

Theorem 1.12. (Lebesgue's Dominated Convergence Theorem) *Let (X,μ) be a measure space. Suppose (f_n) is a sequence in $L^1(X,\mu)$ such that*

$$f(x) = \lim_{n \to \infty} f_n(x)$$

exists almost everywhere on X. If there is a function $g \in L^1(X,\mu)$ such that, for almost all x in X, $n = 1, 2, ...,$

$$|f_n(x)| \le g(x),$$

then $f \in L^1(X,\mu)$ and

$$\lim_{n \to \infty} \int_X |f_n - f| d\mu = 0.$$

If $X = \Omega \subset \mathbf{R}^k$ and if μ is the Lebesgue measure, we write $L^p(\Omega)$ instead of $L^p(X,\mu)$, $1 \le p \le \infty$. If μ is the counting measure on $X = \mathbf{N}$ (or $\mathbf{Z}$), i.e., $\mu(A)$ is the number of elements in A for $A \subset X$, then the corresponding space $L^p(X,\mu)$ is denoted by l^p (or $l^p(\mathbf{Z})$). An element of l^p can be seen as a complex sequence $x = (\xi_n)_{n \ge 1}$ with the norm

$$\|x\|_{l^p} = \left(\sum_{n=1}^{\infty} |\xi_n|^p \right)^{1/p}, \qquad 1 \le p < \infty,$$

$$\|x\|_{l^\infty} = \sup_{n \in \mathbf{N}} |\xi_n|.$$

Similarly, an element of $l^p(\mathbf{Z})$ can be seen as a complex sequence $y = (\xi_n)_{n \in \mathbf{Z}}$ with the norm

$$\|y\|_{l^p} = \left(\sum_{n=-\infty}^{\infty} |\xi_n|^p \right)^{1/p}, \qquad 1 \le p < \infty,$$

$$\|y\|_{l^\infty} = \sup_{n \in \mathbf{Z}} |\xi_n|.$$

1.3.3 Sobolev spaces

Let Ω be a bounded domain in $\mathbf{R}^k$ ($k = 1, 2, ...$). For $\alpha = (\alpha_1, ..., \alpha_k)$, D^α is defined as in Subsection 1.3.1. We denote by $L^p_{loc}(\Omega)$ ($p \ge 1$) the set of Lebesgue measurable functions f on Ω such that $f \in L^p(D)$ for all open subsets D of Ω satisfying $\overline{D} \subset \Omega$. We denote by $C_c^\infty(\Omega)$ the set of infinitely differentiable functions f on Ω such that *supp* $f \subset \Omega$, where *supp* f is the closure of the set of points x of Ω such that $f(x) \neq 0$. Let $u, w \in L^1_{loc}(\Omega)$. Then w is called a *generalized derivative* of u of mixed order α if

$$\int_\Omega u D^\alpha \phi dx = (-1)^{|\alpha|} \int_\Omega w \phi dx \qquad \text{for } \phi \in C_c^\infty(\Omega).$$

We denote w by $D^\alpha u$. The Sobolev space $W^{m,p}(\Omega)$, where m is a positive integer, is the set of all functions u having generalized derivatives up to order m such that

$$D^\alpha u \in L^p(\Omega) \qquad \text{for } |\alpha| \leq m.$$

For $m = 0$, we set $W^{0,p}(\Omega) = L^p(\Omega)$. For $p = 2$, we write $H^m(\Omega)$ for $W^{m,2}(\Omega)$. For $0 \leq p \leq \infty$, $W^{m,p}(\Omega)$ is a Banach space with respect to the norm

$$\|f\|_{W^{m,p}(\Omega)} = \left(\sum_{|\alpha| \leq m} \|D^\alpha f\|^p_{L^p(\Omega)} \right)^{1/p}, \qquad \text{for } 1 \leq p < \infty,$$

$$\|f\|_{W^{m,\infty}(\Omega)} = \max_{|\alpha| \leq m} \|D^\alpha f\|_{L^\infty(\Omega)}.$$

The space $H^m(\Omega)$ is a Hilbert space with respect to the inner product

$$(f,g)_{H^m} = \sum_{|\alpha| \leq m} \int_\Omega D^\alpha f \, \overline{D^\alpha g} dx \qquad \text{for } f,g \in H^m(\Omega).$$

The closure of $C_c^\infty(\Omega)$ in $H^m(\Omega)$ is denoted by $H_0^m(\Omega)$.

For $0 \leq \sigma < 1$, the Sobolev space (of fractional order) $W^{\sigma,p}(\Omega)$, $1 \leq p \leq \infty$, is defined in Chapter 3 (cf. also [Br], p. 196).

Now we state some Sobolev imbedding theorems. Let X, Y be two Banach spaces, $X \subset Y$. The operator $j : X \to Y$ defined by $j(u) = u$ for all $u \in X$ is called the embedding operator of X into Y. One has (cf. [Br], Chap. IX)

Theorem 1.13. *Let Ω be a bounded domain in $\mathbf{R}^k$ such that $\partial\Omega$ is C^1 – smooth.*
   a) If $1 \leq p < k$ then

$$W^{1,p}(\Omega) \subset L^q(\Omega) \qquad \text{for all } q \text{ in } [1, p^*), \quad \frac{1}{p^*} = \frac{1}{p} - \frac{1}{k},$$

   b) If $p = k$ then

$$W^{1,p}(\Omega) \subset L^q(\Omega) \qquad \text{for all } q \in [1, \infty),$$

   c) If $p > k$ then

$$W^{1,p}(\Omega) \subset C(\overline{\Omega}),$$

   where the corresponding embedding operators in a)-c) are compact.

1.4 Analytic functions and harmonic functions

Let Ω be a domain (i.e. an open, connected subset) of the complex plane $\mathbf{C}$ and let f be a complex function defined on Ω. We say that f is *analytic* at $z_0 \in \mathbf{C}$ if

$$\lim_{z \to z_0} \frac{f(z) - f(z_0)}{z - z_0}$$

exists. The function f is said to be analytic on Ω if it is analytic at each point of Ω. One has (cf. [Ru], Chap. 10)

Theorem 1.14. *Let f be an analytic function on $\Omega \subset \mathbf{C}$ and let $z_0 \in \Omega$, $r > 0$ be such that $B_r(z_0) \subset \Omega$, where $B_r(z_0)$ is the (open) disc of radius r centered at z_0. Then f is representable by the power series*

$$f(z) = \sum_{n=0}^{\infty} a_n (z - z_0)^n \qquad \text{for } z \in B_r(z_0)$$

where

$$a_n = \frac{1}{2\pi i} \int_{\partial B_r(z_0)} \frac{f(\zeta)}{(\zeta - z_0)^{n+1}} d\zeta.$$

Theorem 1.15. (Identity Theorem) *Let f_1, f_2 be analytic functions on a domain $\Omega \subset \mathbf{C}$ such that $f_1(z) = f_2(z)$ on a set of points of Ω with an accumulation point in Ω. Then $f_1 = f_2$ on Ω.*

Let Ω be a domain in $\mathbf{R}^k$, $k \geq 2$. A C^2-function f on Ω is said to be *harmonic* if it satisfies the *Laplace equation*

$$\Delta f = 0 \qquad \text{on } \Omega$$

where

$$\Delta = \frac{\partial^2}{\partial x_1^2} + \frac{\partial^2}{\partial x_2^2} + \ldots + \frac{\partial^2}{\partial x_k^2} \qquad \text{for } x = (x_1, x_2, \ldots, x_k).$$

One has the following result related to the eigenfunctions of the Laplace operator (cf. [Br], Chap. 9)

Theorem 1.16. *Let Ω be a bounded domain in $\mathbf{R}^k$. Then there exists an orthonormal basis (e_n) of $L^2(\Omega)$ and a sequence (λ_n) of positive numbers tending to infinity for $n \to \infty$ such that $e_n \in H_0^1(\Omega) \cap C^\infty(\Omega)$ and that*

$$-\Delta e_n = \lambda_n e_n \qquad \text{on } \Omega.$$

Let Γ_0 be an open subset of $\partial \Omega$. A Cauchy problem for the Laplace equation is one of finding a function u satisfying

$$\Delta u = 0 \qquad \text{on } \Omega \tag{1.1}$$

subject to the conditions

$$u|_{\Gamma_0} = f_0, \qquad \left.\frac{\partial u}{\partial n}\right|_{\Gamma_0} = f_1 \tag{1.2}$$

where $\partial/\partial n$ is the normal derivative to Γ_0. We say that u is a weak solution of (1.1)-(1.2) if u satisfies

$$\int_\Omega u\Delta\phi dx = \int_{\Gamma_0} f_0 \frac{\partial\phi}{\partial n} d\sigma - \int_\Omega f_1\phi dx \tag{1.3}$$

for all $\phi \in C^2(\overline{\Omega})$, $\phi = 0$ on a neighborhood of $\partial\Omega \setminus \Gamma_0$, where $d\sigma$ is the surface area element of $\partial\Omega$. The problem (1.1)-(1.2), under some appropriate assumptions, has at most one solution. In fact, one has

Theorem 1.17. *Let Γ_0 be C^1-smooth, let f_0, f_1 be functions in $L^2(\Gamma_0)$. Then (1.1)-(1.2) has at most one weak solution u in $L^2(\Omega)$.*

Proof. For the proof, we rely on the unique continuation property of harmonic functions, according to which, a harmonic function on Ω that is known on an open subset of a domain Ω, is uniquely extendable to a harmonic function on all of Ω (cf. [Pe] where the uniqueness of continuation for solutions of elliptic equations is proved). Indeed, let u_1, u_2 be two weak solutions in $L^2(\Omega)$ of (1.1)-(1.2). We shall prove that $u_1 = u_2$. Putting $w = u_1 - u_2$, one gets in view of (1.3)

$$\int_\Omega w\Delta\phi dx = 0 \tag{1.4}$$

for all ϕ in $C^2(\overline{\Omega})$, $\phi = 0$ on a neighborhood of $\partial\Omega \setminus \Gamma_0$. Let D be a connected component of $\mathbf{R}^k \setminus \overline{\Omega}$ such that $\Omega \cup \Gamma_0 \cup D$ is connected. Put

$$\tilde{w}(x) = w(x) \qquad \text{for } x \text{ in } \Omega,$$
$$= 0 \qquad \text{for } x \text{ in } \Gamma_0 \cup D.$$

From (1.4), we get

$$\int_{\Omega\cup\Gamma_0\cup D} \tilde{w}\Delta\phi dx = 0 \qquad \text{for } \phi \text{ in } C_c^\infty(\Omega \cup \Gamma_0 \cup D).$$

It follows that $\tilde{w}$ is a weak solution of the Laplace equation $\Delta u = 0$ on $\Omega\cup\Gamma_0\cup D$. Since $\tilde{w}$ is in $L^2(\Omega \cup \Gamma_0 \cup D)$, using Theorem 16.1 of [Fr], p. 54, we get $\tilde{w} \in C^\infty(\Omega\cup\Gamma_0\cup D)$. Now, by the unique continuation property of harmonic functions (cf. [Pe]), we get in view of the fact that $\tilde{w} = 0$ on D that $\tilde{w} = 0$ on $\Omega \cup \Gamma_0 \cup D$. It follows that $u_1 = u_2$ on Ω. This completes the proof of Theorem 1.17.

1.5 Fourier transform and Laplace transform

For f in $L^1(\mathbf{R})$, we define the Fourier transform of f by

$$\hat{f}(x) = \int_{-\infty}^{\infty} e^{ixt} f(t)dt.$$

For f in $L^2(\mathbf{R})$, putting for $\epsilon > 0$,

$$F_f(x, \epsilon) = \int_{-\infty}^{\infty} e^{-\epsilon|t|} e^{ixt} f(t)dt,$$

we define the Fourier transform of f as

$$\hat{f}(x) = \lim_{\epsilon \downarrow 0} F_f(x, \epsilon).$$

One has (cf. [St])

Theorem 1.18. *One can associate to each f in $L^2(\mathbf{R})$ a function $\hat{f}$ in $L^2(\mathbf{R})$ so that*
  a) If f is in $L^1(\mathbf{R}) \cap L^2(\mathbf{R})$ then

$$\hat{f}(x) = \int_{-\infty}^{\infty} e^{ixt} f(t)dt,$$

  b) For f in $L^2(\mathbf{R})$ one has

$$\|f\|_{L^2(\mathbf{R})} = \frac{1}{\sqrt{2\pi}} \|\hat{f}\|_{L^2(\mathbf{R})}$$

and

$$\int_{-\infty}^{\infty} f(t)\overline{g(t)}dt = \frac{1}{2\pi} \int_{-\infty}^{\infty} \hat{f}(x)\overline{\hat{g}(x)}dx,$$

  c) One has

$$\lim_{\epsilon \downarrow 0} \|\hat{f} - F_f(., \epsilon)\|_{L^2(\mathbf{R})} = 0$$

$$\lim_{\delta \downarrow 0} \|f - G_{\hat{f}}(., \delta)\|_{L^2(\mathbf{R})} = 0$$

where

$$G_{\hat{f}}(t, \delta) = \frac{1}{2\pi} \int_{-\infty}^{\infty} e^{-\delta|x|} e^{-ixt} \hat{f}(x)dx,$$

  d) For f, g in $L^2(\mathbf{R}^2)$, one has $\widehat{f * g} = \hat{f}\hat{g}$ where

$$(f * g)(t) = \int_{-\infty}^{\infty} f(\tau)g(t - \tau)d\tau.$$

Similarly, we define the *n-dimensional Fourier transform*. In fact, if u is in $L^2(\mathbf{R}^k)$, the Fourier transform $\hat{u}$ in $L^2(\mathbf{R}^k)$ is defined by

$$\hat{u}(y) = \int_{\mathbf{R}^k} e^{ix\cdot y} u(x) dx,$$

$$x.y = x_1 y_1 + \ldots + x_k y_k \qquad \text{for } x = (x_1, \ldots, x_k), y = (y_1, \ldots, y_k).$$

the latter integral converging in the L^2-sense. Similar to the one-dimensional case, we have the inversion formula

$$u(x) = \frac{1}{(2\pi)^k} \int_{\mathbf{R}^k} e^{-ix\cdot y} \hat{u}(y) dy$$

and

$$\|f\|_{L^2(\mathbf{R}^k)} = \frac{1}{\sqrt{(2\pi)^k}} \|\hat{f}\|_{L^2(\mathbf{R}^k)}.$$

We denote by $\mathbf{R}_+$ the set $(0, \infty)$. If f is a function in $L^2(\mathbf{R}_+)$, we define the Laplace transform of f by the integral

$$g(s) = \mathcal{L}f(s) = \int_0^\infty e^{-st} f(t) dt.$$

Upon setting $s = x + iy$ in the latter equality, we get

$$g(x + iy) = \int_0^\infty \left(e^{-xt} f(t) \right) e^{-iyt} dt.$$

Hence, we may think of the Laplace transforms as a Fourier transform. The following theorems are due to Paley and Wiener (cf. [Ru], Chap. 19)

Theorem 1.19. *Suppose g is an analytic function on the half plane $Im\ z > 0$ and let*

$$\sup_{0 < y < \infty} \frac{1}{2\pi} \int_{-\infty}^\infty |g(x + iy)|^2 dx = C < \infty.$$

Then there exists an f in $L^2(\mathbf{R}_+)$ such that

$$g(z) = \int_0^\infty f(t) e^{itz} dt \qquad \text{for } Im\ z > 0$$

and

$$\int_0^\infty |f(t)|^2 dt = C.$$

Theorem 1.20. *Suppose A and C are positive constants and g is an entire function (i.e. analytic on $\mathbf{C}$) such that*

$$|g(z)| \leq Ce^{A|z|} \qquad \text{for all } z \in \mathbf{C}$$

and

$$\int_{-\infty}^{\infty} |g(x)|^2 dx < \infty.$$

Then there exists an f in $L^2(-A, A)$ such that

$$g(z) = \int_{-A}^{A} f(t)e^{itz} dt.$$

2 Regularization of moment problems by truncated expansion and by the Tikhonov method

In the Introduction of the book, we have defined the concept of moment problem in a rather general setting (cf. (0.1)). In this chapter, we shall consider moment problems of the conventional form (cf. (0.3)):

(MP) Find a function u on a domain $\Omega \subset \mathbf{R}^d$ satisfying the sequence of equations

$$\int_\Omega u(x)g_n(x)dx = \mu_n, \qquad n \in \mathbf{N}, \tag{2.1}$$

where (g_n) is a given sequence of functions lying in $L^2(\Omega)$.

The Hausdorff moment problem is a classical example of a moment problem:

Find a function u on (a,b) such that

$$\int_a^b u(x)x^k dx = \mu_k, \qquad k \in \mathbf{N}. \tag{2.2}$$

We also have moment problems of the form

$$\int_0^T u(t)e^{2i\omega_j t}dt = c_{2j-1},$$

$$\int_0^T u(t)e^{-2i\omega_j t}dt = c_{2j}, \tag{2.3}$$

where $i = \sqrt{-1}$ and (ω_j), $j \in \mathbf{N}$, is a sequence of real numbers.

Moment problems of the form (2.3) are called trigonometric moment problems, they occur in the theory of control and are discussed in Krabs' monograph [Kr] (cf. also [AGl] for more general trigonometric moment problems). Hausdorff moment problems occupy a central place in Analysis and in the Applied Sciences, they will be discussed in Chapter 4. Other examples of moment problems will be given in Chapters 5, 6, 7.

Moment problems are usually ill-posed in the sense that they have no solution and that in the case of existence of solutions, there is no continuous dependence on the given data. The present chapter is devoted to some methods of constructing regularized solutions, that is, approximate solutions stable with respect to the given data. Such a construction process is called *regularization*. There are various methods of regularization. Two of them will be considered in this chapter: the *method of truncated expansion* and the *method of Tikhonov* (also called the Tikhonov-Phillips method).

The method of truncated expansion consists in approximating (2.1) by finite moment problems

$$\int_{\Omega} u(y)g_j(y)dy = \mu_j, \qquad j = 1, .., n. \tag{2.4}$$

Solved in the subspace $< g_1, .., g_n >$ generated by $g_1, ..., g_n$, (2.4) is stable. Moreover, with $n \in \mathbf{N}$ chosen appropriately, solutions of (2.4) approximate those of the original problem (2.1). Considering the case where the data $\mu = (\mu_1, .., \mu_n)$ are inexact, we derive some convergence theorems and error estimates for the regularized solutions.

In the method of Tikhonov, we write (2.1) in the form

$$Au = \mu$$

with

$$Au = (\int_{\Omega} ug_1, \int_{\Omega} ug_2, ...), \quad \mu = (\mu_1, \mu_2, ...)$$

and regularize it by the problem of finding a function $u \in L^2(\Omega)$ satisfying the variational equation

$$\beta(u, v)_{L^2(\Omega)} + (Au, Av)_{l^2} = (\mu, Av)_{l^2}, \quad \forall v \in L^2(\Omega) \tag{2.5}$$

$((.,.)_{L^2(\Omega)}$ and $(.,.)_{l^2}$ are respectively the usual inner products on $L^2(\Omega)$ and l^2, and $\beta > 0$). With some smoothness conditions on the solutions u of (2.1), we derive some error estimates for the approximate solutions. Furthermore, we also prove a convergence result when no further assumptions are made on u. This method applies to more general moment problems, such as moment problems in $L^\alpha(\Omega)$ $(1 < \alpha < \infty)$. We note that another regularization method, based on the Backus-Gilbert approach, will be presented in chapter 3. It is similar to the first method of this chapter, except that, instead of using an orthonormal system generated by $g_1, g_2, ...$, the approximate solutions are constructed as combinations of some predetermined basis functions (called the Backus-Gilbert basis functions). This leads to different results, with different conditions and error estimates.

The remainder of the present chapter is divided into two sections. The first method is studied in Section 2.1 which consists of three subsections 2.1.1, 2.1.2, 2.1.3. In Subsection 2.1.1, we shall give a construction of regularized solutions using the method of orthogonal projection in Hilbert space. Subsection 2.1.2 is devoted to the regularization results and error estimates in case of L^2-exact solutions and of H^m-exact solutions. Another error estimate, based on results concerning eigenvalues of the Laplacian, is considered in Section 2.1.3. Section 2.2 consisting of three subsections 2.2.1, 2.2.2, 2.2.3 is devoted to the second regularization method. A convergence result for L^{α^*}-solutions $(1 < \alpha^* < \infty)$ of Problem (2.1) is shown in Subsection 2.2.2 while error estimates between regularized solutions and exact solutions in the $L^2(\Omega)$ case and in the $H^1(\Omega)$ case are derived in Subsections 2.2.1, 2.2.3 respectively. In Subsection 2.2.2, a Banach space version of the Tikhonov method is presented.

2.1 Method of truncated expansion

This regularization method is based on the properties of orthogonal projections in the real $L^2(\Omega)$. Throughout this Section 2.1 we work in the *real $L^2(\Omega)$*.

2.1.1 A construction of regularized solutions

In the sequel, unless stated otherwise, we assume that $g_1, g_2, ...$ are linearly independent (in the algebraic sense) and that the vector space generated by $g_1, g_2, ...$ is dense in $L^2(\Omega)$. We denote by $\| \, . \, \|$ and $(.,.)$ the usual norm and inner product of $L^2(\Omega)$.

Let $\{e_1, e_2, ...\}$ be the orthonormal system constructed from $\{g_1, g_2, \, ...\}$ by the Gram-Schmidt orthogonalization method as follows

$$e_1 = g_1/\|g_1\|$$

$$e_n = \|g_n - \sum_{j=1}^{n-1}(g_n, e_j)e_j\|^{-1}\left(g_n - \sum_{j=1}^{n-1}(g_n, e_j)e_j\right) \qquad (2.6)$$

$$(n = 2, 3, ...).$$

Then $\{e_1, e_2, ...\}$ is an orthonormal basis of $L^2(\Omega)$ and moreover, there exist unique constants C_{ij}, $M_{ij} \in \mathbf{R}$ $(i, j \in N)$ such that $C_{ij} = M_{ij} = 0$ if $i < j$, and that

$$e_i = \sum_{j=1}^{i} C_{ij} g_j, \quad g_i = \sum_{j=1}^{i} M_{ij} e_j, \qquad \forall i \in N.$$

Hence, if we consider the matrices

$$C_n = [C_{ij}]_{1 \le i,j \le n}, \quad M_n = [M_{ij}]_{1 \le i,j \le n},$$

then $M_n = C_n^{-1}$.

We can calculate C_{ij}, M_{ij} as follows.

From (2.6), we have $C_{11} = \|g_1\|^{-1}$. Suppose C_{jk}, $j = 1, ..., n-1$, $k = 1, .., j$ are known. Then

$$C_{nk} = \left(\sum_{j,l=1}^{n} \bar{C}_{nj}\bar{C}_{nl}(g_j, g_l)\right)^{-1/2} \bar{C}_{nk} \qquad (1 \le k \le n)$$

with

$$\bar{C}_{nk} = -\sum_{j=k}^{n-1}\left(\sum_{l=1}^{j} C_{jl}(g_n, g_l)\right) C_{jk} \qquad (1 \le k \le n-1),$$

$$\bar{C}_{nn} = 1.$$

The M_{ij}'s are calculated via the C_{ij}'s by

$$M_{ij} = (g_i, e_j) = \sum_{k=1}^{j} C_{jk}(g_i, g_k), \qquad i, j \in N.$$

Now, we associate to each sequence of real numbers $\mu = (\mu_1, \mu_2, ...)$ the sequence $\lambda = \lambda(\mu) = (\lambda_1, \lambda_2, ...)$ defined by

$$\lambda_i = \lambda_i(\mu) = \sum_{j=1}^{i} C_{ij}\mu_j, \qquad i = 1, 2, ... \tag{2.7}$$

(If μ is a finite sequence, $\mu = (\mu_1, \mu_2, ..., \mu_n)$ then $\lambda = \lambda(\mu)$ is also a finite sequence with the λ_i's $(i = 1, .., n)$ given by (2.7)).

The following result characterizes the solution of minimal norm of the finite moment problem (2.4).

Proposition 2.1. *Let $\mu = (\mu_1, \mu_2, ..)$ be a sequence of real numbers, let $u \in L^2(\Omega)$, and $n \in N$. Then, the following statements are equivalent*

(i) $u \in\ <g_1, .., g_n>$ (the linear space generated by $\{g_1, .., g_n\}$) and

$$(u, g_j) = \mu_j, \qquad 1 \le j \le n. \tag{2.8}$$

(ii) u satisfies (2.8) and

$$\|u\| = \min\{\|v\| : v \in L^2(\Omega) \ and \ (v, g_j) = \mu_j \ for \ 1 \le j \le n\}.$$

(iii)

$$u = \sum_{i=1}^{n} \lambda_i(\mu)e_i.$$

Proof.

$(i) \Leftrightarrow (iii)$. Let $u \in\ <g_1, .., g_n>$, $u = \sum_{j=1}^{n} \alpha_j e_j$. Then

$$(u, g_i) = \sum_{j=1}^{n} \alpha_j(e_j, g_i) = \sum_{j=1}^{n} \alpha_j M_{ij} \quad 1 \le i \le n,$$

i.e.,

$$((u, g_1), ..., (u, g_n))^T = M_n(\alpha_1, ..., \alpha_n)^T.$$

Hence, u satisfies (2.8) $\Leftrightarrow$

$\Leftrightarrow M_n(\alpha_1, ..., \alpha_n)^T = (\mu_1, ..., \mu_n)^T$

$\Leftrightarrow (\alpha_1, ..., \alpha_n)^T = C_n(\mu_1, ..., \mu_n)^T = (\lambda_1, ..., \lambda_n)^T$

$\Leftrightarrow u = \sum_{i=1}^{n} \lambda_i e_i$.

$(ii) \Leftrightarrow (iii)$. Suppose u satisfies (ii). Consider the decomposition $u = v + w$ where $v \in\ <g_1, ..., g_n>$, $w \in\ <g_1, ..., g_n>^{\perp}$. One has

$$(v, g_i) = (u, g_i) = \mu_i, \qquad 1 \le i \le n.$$

Hence, by the above proof, $v = \sum_{i=1}^{n} \lambda_i e_i$. On the other hand, $\|v\| \le \|u\|$, and then $\|v\| = \|u\|$ (by (ii)). Thus

$$\|w\|^2 = \|u\|^2 - \|v\|^2 = 0,$$

i.e. $u = v = \sum_{i=1}^{n} \lambda_i e_i$. We have (iii).

Conversely, suppose (iii) holds. Let $v \in L^2(\Omega)$ be such that $(v, g_j) = \mu_j$, $j = 1, .., n$. Let Pv be the orthogonal projection of v on $< g_1, .., g_n >$. Then, as above, we have

$$(Pv, g_j) = (v, g_j) = \mu_j, \qquad 1 \le j \le n,$$

i.e., Pv satisfies (i). Hence,

$$Pv = \sum_{i=1}^{n} \lambda_i e_i = u, \quad \|u\| = \|Pv\| \le \|v\|,$$

i.e., (ii) holds for u. Our proof is completed.

Let μ be a sequence of real numbers. For each $n = 1, 2, ..$, we denote by $p^n = p^n(\mu)$ the unique element of $L^2(\Omega)$ that satisfies the equivalent conditions in Proposition 2.1. Remark that p^n can be defined by (2.7) and (iii), Proposition 2.1 or by (i), Proposition 2.1. Indeed, by putting

$$p^n(\mu) = \sum_{i=1}^{n} \xi_i g_i,$$

we have, from (2.8),

$$\sum_{i=1}^{n} \xi_i(g_i, g_j) = \mu_j, \qquad j = 1, 2, ..., n.$$

This is a Cramer system of linear equations. By the linear independence of $g_1, ..., g_n$, we have

$$\det((g_i, g_j))_{1 \le i, j \le n} \ne 0.$$

Hence $\xi_1, ..., \xi_n$ are determined uniquely.

In fact, one has

$$\xi^T = G_n^{-1} \mu^T$$

for $\xi = (\xi_1, ..., \xi_n)$, $\mu = (\mu_1, ..., \mu_n)$, $G_n = [(g_i, g_j)]_{1 \le i, j \le n}$, where A^T is the transpose of the matrix A. We note that the Gram matrix $G_n = [(g_i, g_j)]_{1 \le i, j \le n}$ in the above linear system may be ill-conditioned, depending on the g_i's. Now, if the g_i's are near-orthogonal then G_n is well-conditioned (the condition number of G_n is 1 if $\{g_i : i \in \mathbf{N}\}$ is an orthogonal sequence). However, G_n may be severely ill-conditioned, in general. For example, in the Hausdorff moment problem on $\Omega = (0, 1)$, the Gram matrices are segments of the Hilbert matrix $((g_i, g_j) = (i + j - 1)^{-1}, \forall i, j \in \mathbf{N})$ and the condition numbers are very large. In fact, as proved in Section 3 of [Tay1], the condition number $P(G_n)$ of G_n satisfies

$$P(G_n) \ge \frac{[(2n)!]^2}{(n!)^4} \sim \frac{16^n}{\pi n}.$$

We also refer to this paper and the references therein for many interesting issues related to Gram matrices and their condition numbers. In the case G_n is ill-conditioned, various regularization methods in numerical linear algebra can be used to find stabilized approximate solutions for $p^n(\mu)$. However, we will not get into the numerical calculations of $p^n(\mu)$ here.

The continuous dependence of $p^n(\mu)$ on μ is shown in the following

Proposition 2.2. *Let $\mu^m = (\mu_j^m)$, $\mu = (\mu_j)$, $m=1,2,\ldots$, be sequences of real numbers such that*

$$\mu_j^m \longrightarrow \mu_j \quad as \ m \longrightarrow \infty, \quad \forall j \in \mathbf{N}.$$

Then, for each $n \in \mathbf{N}$ fixed,

$$p^n(\mu^m) \longrightarrow p^n(\mu) \quad in \ L^2(\Omega) \ as \ m \to \infty.$$

This proposition is a direct consequence of the representation in (iii), Proposition 2.1.

2.1.2 Convergence of regularized solutions and error estimates

In what follows, we shall occasionally assume the following approximation property of $V_n = < g_1, .., g_n >$ to $L^2(\Omega)$:

For each $m \in N$, there exists $C = C(m, \Omega) > 0$ such that for all $v \in H^m(\Omega)$,

$$\|v - P_{V_n}v\| \leq C\|v\|_{H^m(\Omega)}n^{-m}, \quad n = 1, 2, \ldots. \tag{2.9}$$

Here $H^m(\Omega)$, $m = 1, 2, ..$ are the usual Sobolev spaces on Ω and P_{V_n} denotes the orthogonal projection of $L^2(\Omega)$ onto V_n. Note that (2.9) is satisfied if the V_n's are spaces of polynomials or finite element spaces (see e.g. [Ci]).

We now prove the convergence of the approximate solutions $p^n(\mu)$ in the case of *exact data*.

Theorem 2.3. *Let $\mu = (\mu_j)$ be a sequence of real numbers. Then*
(i) (2.1) has at most one solution in $L^2(\Omega)$.
(ii) A necessary and sufficient condition for the existence of a solution of (2.1) is that

$$\sum_{i=1}^{\infty} \left(\sum_{j=1}^{i} C_{ij}\mu_j \right)^2 < \infty. \tag{2.10}$$

(iii) If u is the (unique) solution of (2.1) then

$$p^n(\mu) \longrightarrow u \quad in \ L^2(\Omega) \ as \ n \to \infty.$$

Moreover, if $u \in H^m(\Omega)$ and (2.9) holds, then

$$\|p^n(\mu) - u\| \leq C\|u\|_{H^m(\Omega)}n^{-m}, \quad \forall n \in \mathbf{N}. \tag{2.11}$$

Proof. (i) is obvious since $< g_1, g_2, \ldots >$ is dense in $L^2(\Omega)$.

(ii) Let u be a solution of (2.1). We have, for $i \in \mathbf{N}$,

$$(u, e_i) = \left(u, \sum_{j=1}^{i} C_{ij} g_j \right) = \sum_{j=1}^{i} C_{ij} \mu_j.$$

Hence,

$$\sum_{i=1}^{\infty} \left(\sum_{j=1}^{i} C_{ij} \mu_j \right)^2 = \sum_{i=1}^{\infty} (u, e_i)^2 = \|u\|^2 < \infty,$$

i.e., (2.12) holds. Conversely, if (2.12) holds, then we see, by direct computation, that

$$u = \sum_{i=1}^{\infty} \left(\sum_{j=1}^{i} C_{ij} \mu_j \right) e_i \qquad (2.12)$$

belongs to $L^2(\Omega)$ and is a solution of (2.1).

(iii) Let $u \in L^2(\Omega)$ be a solution of (2.1). Then u is, by (ii), of the form (2.12). Hence, from (2.7),

$$u = \sum_{i=1}^{\infty} \lambda_i(\mu) e_i.$$

By Proposition 2.1, (iii), we have, for $n \in N$

$$p^n(\mu) = P_{V_n}(u). \qquad (2.13)$$

The convergence of $p^n(\mu)$ to u (in $L^2(\Omega)$) thus follows from a well-known property of Hilbert spaces.

Inequality (2.11) is a direct consequence of (2.13) and (2.9). This completes the proof of Theorem 2.3.

The following theorem shows that in the case of *inexact data*, solutions of the finite moment problems (2.8) are stabilized approximations of those of the original problem (2.1). As usual, we denote by $\|\mu\|_\infty$ the sup-norm of the element $\mu = (\mu_i)$ of l^∞

$$\|\mu\|_\infty = \sup\{|\mu_i| : i \in \mathbf{N}\}.$$

Theorem 2.4. *Let $u_0 \in L^2(\Omega)$ be the solution of (2.1) corresponding to $\mu^0 = (\mu_1^0, \mu_2^0, ..)$. For $0 < \epsilon < \|g_1\|^2$, let*

$$n(\epsilon) = [f^{-1}(\epsilon^{-1/2})] \quad (f \text{ is given by } (2.17), (2.18)). \qquad (2.14)$$

Then there exists a function $\eta(\epsilon)$ such that $\lim_{\epsilon \to 0} \eta(\epsilon) = 0$ and that for all sequences μ satisfying

$$\|\mu - \mu^0\|_\infty \le \epsilon,$$

we have

$$\|p^{n(\epsilon)}(\mu) - u_0\| \leq \eta(\epsilon).$$

Moreover, if (2.9) holds and if $u_0 \in H^m(\Omega)$, then

$$\|p^{n(\epsilon)}(\mu) - u_0\| \leq \epsilon^{1/2} + \frac{C\|u_0\|_{H^m(\Omega)}}{(n(\epsilon))^m}.$$

Proof. We have

$$\|p^n(\mu) - u_0\| \leq \|p^n(\mu) - p^n(\mu^0)\| + \|p^n(\mu^0) - u_0\| \qquad (2.15)$$

We denote by $\| \cdot \|_2$ and $\| \cdot \|_\infty$ respectively the Euclidean norm and the sup-norm in $\mathbf{R}^n$, and by $\| C_n \|$ the norm of $C_n = [C_{ij}]_{i,j=1,2,..,n}$, regarded as a linear mapping of $(\mathbf{R}^n, \| \cdot \|_\infty)$ to $(\mathbf{R}^n, \| \cdot \|_2)$:

$$\| C_n \| = \sup\{\| C_n v^T \|_2 : v \in \mathbf{R}^n, \| v \|_\infty \leq 1\}.$$

Hence

$$
\begin{aligned}
\|p^n(\mu) - p^n(\mu^0)\| &= \|\sum_{i=1}^{n}(\lambda_i(\mu) - \lambda_i(\mu^0))e_i\| \\
&= \|(\lambda_1(\mu) - \lambda_1(\mu^0), .., \lambda_n(\mu) - \lambda_n(\mu^0))\| \\
&= \|C_n(\mu_1 - \mu_1^0, .., \mu_n - \mu_n^0)^T\| \\
&\leq \|C_n\|\|(\mu_1 - \mu_1^0, .., \mu_n - \mu_n^0)^T\|_\infty \\
&\leq \|C_n\| \|\mu - \mu^0\|_\infty. \qquad (2.16)
\end{aligned}
$$

Let f be a strictly increasing function of $[1, \infty)$ onto $[\|g_1\|^{-1}, \infty)$ such that

$$\|C_n\| \leq f(n), \qquad \forall n \geq 1. \qquad (2.17)$$

We can choose, for example,

$$f(t) = t - 1 + (t - [t])\|C_{[t]+1}\| + \sum_{j=1}^{[t]} \|C_j\|, \qquad (2.18)$$

in which,

$$f(1) = \|C_1\| = \|g_1\|^{-1}, \; f(n) = n - 1 + \sum_{j=1}^{n} \|C_j\|,$$

and f is affine in each interval $[n, n+1]$, $n \in \mathbf{N}$.

For $0 < \epsilon < \|g_1\|^2$, we have $\epsilon^{-1/2} > \|g_1\|^{-1}$ and $f^{-1}(\epsilon^{-1/2}) \geq 1$. With $n(\epsilon)$ given by (2.14), $n(\epsilon) \leq f^{-1}(\epsilon^{-1/2})$ and thus

$$\|C_{n(\epsilon)}\| \leq f(n(\epsilon)) \leq \epsilon^{1/2}.$$

This and (2.16) imply

$$\|p^{n(\epsilon)}(\mu) - p^{n(\epsilon)}(\mu^0)\| \leq \epsilon^{1/2}. \qquad (2.19)$$

On the other hand,

$$\|p^{n(\epsilon)}(\mu^0) - u_0\| = \left(\sum_{j=n(\epsilon+1)}^{\infty} (\lambda_j(\mu^0))^2 \right)^{1/2}. \tag{2.20}$$

Put

$$\eta(\epsilon) = \epsilon^{1/2} + \left(\sum_{j=n(\epsilon)+1}^{\infty} \left(\sum_{i=1}^{j} C_{ji}\mu_i^0 \right)^2 \right)^{1/2}.$$

We have from (2.15), (2.19), (2.20),

$$\|p^{n(\epsilon)}(\mu) - u_0\| \le \eta(\epsilon), \qquad 0 < \epsilon < \|g_1\|^2.$$

As $\epsilon \to 0$, we have $f^{-1}(\epsilon^{-1/2}) \to \infty$ and $n(\epsilon) \to \infty$. By Theorem 2.3,

$$\sum_{j=n(\epsilon)+1}^{\infty} \left(\sum_{i=1}^{j} C_{ji}\mu_i^0 \right)^2 \to 0.$$

Thus $\eta(\epsilon) \to 0$ as $\epsilon \to 0$.

Now suppose $u_0 \in H^m(\Omega)$ and that (2.9) holds. By Theorem 2.3,

$$\|p^n(\mu^0) - u_0\| \le C\|u_0\|_{H^m(\Omega)} n^{-m}.$$

Together with (2.15), (2.19), this gives

$$\|p^{n(\epsilon)}(\mu) - u_0\| \le \epsilon^{1/2} + C\|u_0\|_{H^m(\Omega)}(n(\epsilon))^{-m}.$$

Our proof is completed.

It would be interesting to give some estimates of $\|C_n\|$ and a more explicit form of $f(n)$ (in the statement of Theorem 2.4) using the condition number of the Gram matrix $[(g_i, g_j)]_{1 \le i,j \le n}$. We have the representation

$$g_i = \sum_{j=1}^{i} M_{ij}e_j,$$

where, we recall, $M_{ij} = 0$ for $i < j$ and

$$M_n = C_n^{-1} \quad \text{for} \quad M_n = [M_{ij}]_{1 \le i,j \le n}.$$

Hence, using the orthonormality of (e_i) one has

$$(g_i, g_j) = \sum_{k=1}^{n} M_{ik}M_{jk}.$$

It follows that

$$G_n = M_n M_n^T, \quad G_n^{-1} = C_n^T C_n.$$

The condition number of G_n is

$$P(G_n) = \lambda_{max}(G_n)\lambda_{max}(G_n^{-1})$$

where $\lambda_{max}(A)$ denotes the largest eigenvalue of the (symmetric, positive definite) matrix A. One has

$$\begin{aligned}
\lambda_{max}(G_n) &= \sup_{\|x\|_2 \le 1} (G_n x^T, x^T)\\
&= \sup_{\|x\|_2 \le 1} (M_n M_n^T x^T, x^T)\\
&= \sup_{\|x\|_2 \le 1} (M_n^T x^T, M_n^T x^T) = \|M_n^T\|_2^2
\end{aligned}$$

where $\|x\|_2$ denotes the Euclidean norm of $x \in \mathbf{R}^n$ and $\|A\|_2$ denotes the norm of the matrix A regarded as a linear operator on $\mathbf{R}^n$ with the Euclidean norm.

Similary

$$\lambda_{max}(G_n^{-1}) = \|C_n\|_2^2.$$

It follows that

$$P(G_n) = \|M_n^T\|_2^2 \|C_n\|_2^2.$$

We recall that, in Theorem 2.4,

$$\|C_n\| = \sup\{\|C_n v^T\|_2 : v \in \mathbf{R}^n, \ \|v\|_\infty \le 1\}.$$

By direct computation, one has

$$\|C_n v^T\|_2 \le \|C_n\|_2 \|v\|_2 \le \|C_n\|_2 \sqrt{n}\|v\|_\infty$$

which implies

$$\|C_n\| \le \sqrt{n}\|C_n\|_2 \le \frac{\sqrt{n}}{\|M_n^T\|_2}\sqrt{P(G_n)}.$$

Hence, we have a relation between $\|C_n\|$ and the condition number of the Gram matrix G_n.

Let f be a function giving the dimension of the approximation as in Theorem 2.4 (cf.(2.14)). We shall derive a simplified expression for $f(n)$. To this end, we first give an estimate of $\|M_n^T\|_2$. For $v = (1, 0, ..., 0) \in \mathbf{R}^n$, one has

$$M_n^T v^T = (M_{11}, 0, ..., 0)^T = (\|g_1\|, 0, ..., 0)^T.$$

Hence

$$\|g_1\| \le \|M_n^T v^T\|_2 \le \|M_n^T\|_2 \|v^T\|_2 = \|M_n^T\|_2$$

and

$$\|C_n\| \le \frac{\sqrt{n}}{\|g_1\|}\sqrt{P(G_n)}.$$

We can therefore choose a function f such that

$$f(n) = \frac{\sqrt{n}}{\|g_1\|}\sqrt{P(G_n)}.$$

2.1.3 Error estimates using eigenvalues of the Laplacian

In this subsection, with other assumptions on u_0, μ^0, we shall get more specific error estimates. The subsection is divided into two parts. In the first part, we shall give some definitions and notations. The second part is devoted to a derivation of error estimates.

Definitions and notations.

We assume in this section that Ω is a bounded domain with smooth boundary. It is known that the eigenvalue problem for the Laplacian:

$$-\Delta u = \alpha u \quad in\ \Omega,$$
$$u = 0 \quad on\ \partial\Omega$$

has an increasing sequence of eigenvalues

$$0 < \alpha_1 \leq \alpha_2 \leq ... \leq \alpha_n \leq ....$$

such that

(i) $\alpha_n \to \infty$ as $n \to \infty$.

(ii) The corresponding sequence of (normalized) eigenfunctions $b_1, b_2, \ldots$ is an orthonormal system of $L^2(\Omega)$.

For $k \in \mathbf{N}$, we put (see e.g. Chap. 3, [Th] or Chap. IV [Mi])

$$\overset{\circ}{H}{}^k(\Omega) = \{v \in L^2(\Omega) : \sum_{m=1}^{\infty} \alpha_m^k(v, b_m)^2 < \infty\}.$$

Then $\overset{\circ}{H}{}^k(\Omega)$ is a Hilbert space with inner product

$$(v, w)_{\overset{\circ}{H}{}^k(\Omega)} = \sum_{m=1}^{\infty} \alpha_m^k(v, b_m)(w, b_m), \quad v, w \in \overset{\circ}{H}{}^k(\Omega)$$

and norm

$$\|v\|_{\overset{\circ}{H}{}^k(\Omega)} = \left[(v, v)_{\overset{\circ}{H}{}^k(\Omega)}\right]^{1/2}.$$

We need the following property of $\overset{\circ}{H}{}^k(\Omega)$.

Lemma 2.1. *Let Ω be a bounded domain in $\mathbf{R}^d$. We have*

$$\overset{\circ}{H}{}^k(\Omega) = \{v \in H^k(\Omega) : \Delta^j v = 0 \text{ on } \partial\Omega \text{ for } j < k/2\}$$

and the norms $\|.\|_{\overset{\circ}{H}{}^k(\Omega)}$ and $\|.\|_{H^k(\Omega)}$ are equivalent on $\overset{\circ}{H}{}^k(\Omega)$. Moreover,

$$\|u\|_{\overset{\circ}{H}{}^k(\Omega)} \leq d^{k/2}|v|_{k,\Omega}, \quad u \in \overset{\circ}{H}{}^k(\Omega), \tag{2.21}$$

with

$$|u|_{k,\Omega} = \left(\sum_{|\gamma|=k} \|D^\gamma u\|^2 \right)^{1/2}.$$

Proof. The first part of this lemma is the content of Lemma 1, Chap. 3, [Th]. The proof of (2.21) is also based on that of the quoted lemma.

For $k = 2p$ even, we have

$$\|u\|_{\overset{\circ}{H^k(\Omega)}} = \|\Delta^p u\|$$

$$= \|\sum_{|\gamma|=p} C_\gamma D^{2\gamma} u\|$$

$$\leq |u|_{2p,\Omega} \sum_{|\gamma|=p} C_\gamma$$

$$= d^{k/2} |u|_{k,\Omega},$$

where

$$C_\gamma = \frac{p!}{k_1!...k_d!} \quad for \ \gamma = (k_1,..,k_d).$$

For $k = 2p + 1$ odd, we have

$$\|u\|^2_{\overset{\circ}{H^k(\Omega)}} = \|\nabla(\Delta^p u)\|^2_{[L^2(\Omega)]^n} = \sum_{i=1}^{d} \left\| \frac{\partial}{\partial x_i}(\Delta^p u) \right\|^2.$$

But by the above proof, for $1 \leq i \leq d$,

$$\left\| \frac{\partial}{\partial x_i}(\Delta^p u) \right\| = \left\| \Delta^p \left(\frac{\partial u}{\partial x_i} \right) \right\| \leq d^p \left| \frac{\partial u}{\partial x_i} \right|_{2p,\Omega} \leq d^p |u|_{2p+1,\Omega}.$$

Hence,

$$\|u\|^2_{\overset{\circ}{H^k(\Omega)}} \leq \sum_{i=1}^{d} d^{2p} |u|^2_{k,\Omega} = d^k |u|^2_{k,\Omega}.$$

This completes the proof of Lemma 2.1.

Error Estimates

Since $e_1, e_2, ..$ is an orthonormal system in $L^2(\Omega)$, there exists a unique linear mapping φ of $L^2(\Omega)$ onto $L^2(\Omega)$ such that $\varphi(e_i) = b_i$, $\forall i \in \mathbf{N}$. It is clear that φ is a Hilbert isomorphism,

$$\varphi(v) = \sum_{i=1}^{\infty} (v, e_i) b_i, \quad \forall v \in L^2(\Omega),$$

and that φ is fully determined by the set of functions $g_1, g_2, ...$ We have the

Theorem 2.5. *Suppose u_0 is a solution of (2.1), corresponding to $\mu^0 = (\mu_1^0, \mu_2^0, ...)$ and that μ^0 satisfies*

$$\sum_{i=1}^{\infty} \alpha_i^k \left(\sum_{j=1}^{i} C_{ij}\mu_j^0 \right)^2 \leq E^2, \tag{2.22}$$

for some $E > 0$, $k \in \mathbf{N}$.

Then, for all sequences μ such that

$$\|\mu - \mu^0\|_{\infty} \leq \epsilon, \qquad \epsilon > 0,$$

we have, with $n(\epsilon)$ as in Theorem 2.4,

$$\|p^{n(\epsilon)}(\mu) - u_0\| \leq \epsilon^{1/2} + E\alpha_{n(\epsilon)+1}^{-k/2}, \tag{2.23}$$

or equivalently

$$\|p^{n(\epsilon)}(\mu) - u_0\| \leq \epsilon^{1/2} + d^{k/2}|\varphi(u_0)|_{k,\Omega}^{1/2}\alpha_{n(\epsilon)+1}^{-k/2}. \tag{2.24}$$

Proof. For $n \in \mathbf{N}$, we decompose $u_0 = v + w$ with $v = P_{V_n}u_0$. Hence

$$u_0 - p^n(\mu) = (v - p^n(\mu)) + w.$$

Since $v - p^n(\mu) \in V_n$, $w \in V_n^{\perp}$, we have

$$\|p^n(\mu) - u_0\| = \left(\|p^n(\mu) - v\|^2 + \|w\|^2 \right)^{1/2}$$
$$\leq \|p^n(\mu) - v\| + \|w\|. \tag{2.25}$$

As in (2.13), we have $v = P_{V_n}u_0 = p^n(\mu^0)$. For $n = n(\epsilon)$,

$$\|p^{n(\epsilon)}(\mu) - v\| = \|p^{n(\epsilon)}(\mu^0) - p^{n(\epsilon)}(\mu)\| \leq \epsilon^{1/2} \tag{2.26}$$

(see (2.19)).

On the other hand, since for $i \in \mathbf{N}$,

$$(\varphi(w), b_i) = (\varphi(w), \varphi(e_i)) = (w, e_i) =$$
$$= \begin{cases} 0 & \text{if } i \leq n \\ (u_0, e_i) & \text{if } i > n \end{cases}$$
$$= \begin{cases} 0 & \text{if } i \leq n \\ (\varphi(u_0), b_i) & \text{if } i > n, \end{cases}$$

one has

$$\|\varphi(w)\|^2 = \sum_{i=1}^{\infty} (\varphi(w), b_i)^2$$
$$= \sum_{i=n+1}^{\infty} (\varphi(u_0), b_i)^2$$
$$\leq \alpha_{n+1}^{-k} \sum_{i=n+1}^{\infty} \alpha_i^k (\varphi(u_0), b_i)^2$$
$$\leq \alpha_{n+1}^{-k} \|\varphi(u_0)\|_{\overset{\circ}{H}^k(\Omega)}^2 \tag{2.27}$$
$$\leq d^k \alpha_{n+1}^{-k} |\varphi(u_0)|_{k,\Omega}^2 \quad \text{(by Lemma 2.1)}. \tag{2.28}$$

But

$$\|\varphi(u_0)\|^2_{\overset{\circ}{H}{}^k(\Omega)} = \sum_{i=1}^{\infty} \alpha_i^k (u_0, e_i)^2$$

$$= \sum_{i=1}^{\infty} \alpha_i^k \left(\sum_{j=1}^{i} C_{ij}(u_0, g_j) \right)^2$$

$$= \sum_{i=1}^{\infty} \alpha_i^k \left(\sum_{j=1}^{i} C_{ij} \mu_j^0 \right)^2 \le E^2.$$

Hence,

$$\|w\| = \|\varphi(w)\| \le E d^{k/2} \alpha_{n+1}^{-k/2}.$$

In view of (2.25), (2.26), (2.28), this completes our proof.

2.2 Method of Tikhonov

This regularization method consists in approximating (2.1) by coercive variational equations. We consider three cases (corresponding to three subsections). In the first case (Subsection 2.2.1), we assume that the exact solutions are in $L^2(\Omega)$. In the second case (Subsection 2.2.2), exact solutions are in $L^{\alpha^*}(\Omega)$, $(1 < \alpha^* < \infty)$. Finally, Subsection 2.2.3 gives error estimates corresponding to the exact solution in $H^1(\Omega)$. As in the preceding section, we also take $L^2(\Omega)$ as the real $L^2(\Omega)$, and likewise we work in the real spaces $L^\alpha(\Omega)$, $L^{\alpha^*}(\Omega)$ and $H^1(\Omega)$.

2.2.1 Case 1: exact solutions in $L^2(\Omega)$

Statement of the problem.

Before presenting our method of regularization, we first remark that (2.1) is equivalent to the problem of finding $u \in L^2(\Omega)$ such that

$$(u, \bar{g}_j) = \bar{\mu}_j, \qquad j = 1, 2, \ldots$$

with

$$\bar{g}_j = 2^{-j} \|g_j\|^{-1} g_j, \ \bar{\mu}_j = 2^{-j} \|g_j\|^{-1} \mu_j, \ \ j = 1, 2, \ldots.$$

Hence, replacing g_j, μ_j by $\bar{g}_j$, $\bar{\mu}_j$, we can assume without loss of generality that

$$\sum_{j=1}^{\infty} \|g_j\|^2 < \infty.$$

Let l^2 be the usual Hilbert space of real sequences $(\mu_1, \mu_2, ..)$ such that

$$\sum_{j=1}^{\infty} |\mu_j|^2 < \infty.$$

For $u \in L^2(\Omega)$, define

$$Au = \left(\int_\Omega u g_1, \int_\Omega u g_2, \dots \right). \tag{2.29}$$

Since

$$\sum_{j=1}^\infty \left(\int_\Omega u g_j \right)^2 \leq \sum_{j=1}^\infty \left(\int_\Omega u^2 \right) \left(\int_\Omega g_j^2 \right)$$

$$= \|u\|^2 \sum_{j=1}^\infty \|g_j\|^2, \quad \forall u \in L^2(\Omega),$$

we see that A is a continuous linear mapping of $L^2(\Omega)$ into l^2 with

$$\|A\|^2 \leq \sum_{j=1}^\infty \|g_j\|^2.$$

The equation (2.1) thus can be written in the form

$$Au = \mu, \qquad \text{with } \mu = (\mu_1, \mu_2, \dots) \in l^2.$$

In what follows (except in Section 2.3.), we do not assume that $< g_1, g_2, \dots >$ is dense in $L^2(\Omega)$. Thus A may not be injective and (2.1) may fail to have a unique solution.

The Problem in a Hilbert space setting.

The following results hold for linear equations in Hilbert spaces. Hence they are stated in an abstract setting:

Let $(X, (.,.), \|.\|)$ and $(Y, (.,.)_Y, \|.\|_Y)$ be (real) Hilbert spaces and let A be a continuous linear mapping from X to Y. For $\mu \in Y$, we consider the equation

$$Au = \mu, \qquad u \in X. \tag{2.30}$$

When A^{-1} does not exist or does exist but is not bounded (this is often the case in moment problems, see "Notes and remarks" of this chapter), this problem is ill-posed. We shall regularize it by considering the following family of coercive variational equations of finding $u \in X$ such that

$$\beta(u, v) + (Au, Av)_Y = (\mu, Av)_Y, \qquad \forall v \in X, \tag{2.31}$$

with $\beta > 0$.

The stable solvability of (2.31) is shown in the following

Proposition 2.6. *For each $\beta > 0$ fixed, (2.31) has a unique solution $u = u^\beta(\mu)$ which depends continuously on $\mu \in Y$.*

This is a direct consequence of the Lax-Milgram theorem (cf. Chapter 1). We only need to remark that

$$\|u^\beta(\mu) - u^\beta(\mu')\| \leq \|A\| \ \beta^{-1} \ \|\mu - \mu'\|_Y, \qquad \mu, \mu' \in Y$$

Now, since $A : X \to Y$ is linear and continuous, we know that the adjoint operator $A^* : Y \to X$ is also linear and continuous. Moreover, A^*A is a positive, self-adjoint operator in X. By Theorem 1.8, there exists a unique positive, self-adjoint operator $C : X \to X$ such that $C^2 = A^*A$. The following result shows that we can regularize (2.30) by the solutions $u^\beta(\mu)$ of (2.31).

Theorem 2.7. *Let $u_0 \in X$, $\mu^0 \in Y$ be such that*

$$Au_0 = \mu^0. \tag{2.32}$$

(a) Suppose $u_0 = A^*\mu_1$ for some $\mu_1 \in Y$. For $\epsilon > 0$, we choose $\beta(\epsilon) = \epsilon$. Then, if $\mu \in Y$ satisfies

$$\|\mu - \mu^0\|_Y \leq \epsilon \tag{2.33}$$

we have

$$\|u^{\beta(\epsilon)}(\mu) - u_0\| \leq \frac{1}{2}(1 + \|\mu_1\|_Y)\epsilon^{1/2}.$$

(b) Suppose $u_0 = Cu_1$ for some $u_1 \in X$. For $\epsilon > 0$, we choose $\beta(\epsilon) = \epsilon^{2/3}$. Then, for all μ satisfying (2.33), we have

$$\|u^{\beta(\epsilon)}(\mu) - u_0\| \leq (\|A\| + \|u_1\|/\sqrt{2})\epsilon^{1/3}.$$

Proof. Equation (2.32) gives

$$(Au_0, Av)_Y = (\mu^0, Av)_Y, \qquad \forall v \in X.$$

Subtracting this from (2.31), and letting $v = u - u_0$ in the equation thus obtained, we get

$$\beta\|u - u_0\|^2 + \|A(u - u_0)\|_Y^2 =$$
$$= (\mu - \mu^0, A(u - u_0)) - \beta(u_0, u - u_0). \tag{2.34}$$

(a) We have in this case

$$RHS \ of \ (2.34) = (\mu - \mu^0, A(u - u_0)) - \beta(\mu_1, A(u - u_0))$$
$$\leq (\|\mu - \mu^0\|_Y + \beta\|\mu_1\|_Y) \|A(u - u_0)\|_Y$$
$$\leq \frac{1}{4} (\|\mu - \mu^0\|_Y + \beta\|\mu_1\|_Y)^2 + \|A(u - u_0)\|_Y^2.$$

For $\beta = \beta(\epsilon) = \epsilon$, one has

$$\|u^\beta(\mu) - u_0\| = \|u - u_0\|$$
$$\leq \frac{1}{2\sqrt{\beta}} (\|\mu - \mu^0\|_Y + \beta\|\mu_1\|_Y)$$
$$\leq \frac{1}{2\sqrt{\epsilon}}(\epsilon + \epsilon\|\mu_1\|_Y)$$
$$= \frac{1}{2}\epsilon^{1/2}(1 + \|\mu_1\|_Y).$$

(b) Noting that

$$\|A(u - u_0)\|_Y^2 = (A^* A(u - u_0), u - u_0)$$
$$= (C^2(u - u_0), u - u_0) = \|C(u - u_0)\|^2,$$

we have, from (2.34)

$$\beta\|u - u_0\|^2 + \|C(u - u_0)\|^2 \leq$$
$$\leq \|A\| \|\mu - \mu^0\|_Y \|u - u_0\| + \beta|(u_1, C(u - u_0))|$$
$$\leq \frac{\beta}{2}\|u - u_0\|^2 + \frac{1}{2\beta}\|A\|^2\|\mu - \mu^0\|_Y^2 +$$
$$+ \frac{\beta^2}{4}\|u_1\|^2 + \|C(u - u_0)\|^2.$$

Thus

$$\frac{\beta}{2}\|u - u_0\|^2 \leq \frac{1}{2\beta}\|A\|^2\|\mu - \mu^0\|_Y^2 + \frac{\beta^2}{4}\|u_1\|^2,$$

i.e.,

$$\|u - u_0\| \leq \left(\frac{1}{2\beta^2}\|A\|^2\|\mu - \mu^0\|_Y^2 + \frac{\beta}{2}\|u_1\|^2\right)^{1/2}$$
$$\leq \left(\|A\|^2\epsilon^{2/3} + \frac{1}{2}\|u_1\|^2\epsilon^{2/3}\right)^{1/2}$$
$$\leq \left(\|A\| + \frac{1}{\sqrt{2}}\|u_1\|\right)\epsilon^{1/3}.$$

This completes the proof of Theorem 2.7.

The Moment Problem in an $L^2(\Omega)$-setting.

We now consider the particular case of the moment problem (2.1), i.e., when A is given by (2.29), $X = L^2(\Omega)$, $Y = l^2$.

In this case, A^* is given by

$$A^*\mu = \sum_{j=1}^\infty \mu_j g_j, \qquad \mu \in l^2. \tag{2.35}$$

In fact, one can check that the latter series is convergent in $L^2(\Omega)$. Moreover, for $u \in L^2(\Omega)$,

$$\left(\sum_{j=1}^\infty \mu_j g_j, u\right) = \sum_{j=1}^\infty \mu_j (Au)_j$$
$$= (\mu, Au)_{l^2} = (A^*\mu, u)$$

which proves (2.35).

The relation (2.35) implies that

$$A^*Au = \sum_{j=1}^{\infty}(u, g_j)g_j.$$

Hence (2.31) becomes

$$\beta\int_{\Omega}uv + \sum_{j=1}^{\infty}\left(\int_{\Omega}ug_j\right)\left(\int_{\Omega}vg_j\right) = \sum_{j=1}^{\infty}\mu_j\left(\int_{\Omega}vg_j\right), \qquad \forall v \in L^2(\Omega),$$

or in equivalent form

$$\beta u + \sum_{j=1}^{\infty}\left(\int_{\Omega}ug_j\right)g_j = \sum_{j=1}^{\infty}\mu_jg_j \qquad a.e. \ on \ \Omega.$$

We now check that the condition $A^*\mu_1 = u_0$ in Theorem 2.7 is equivalent to the following:

$$\mu_i^0 = \sum_{j=1}^{\infty}\mu_j^1(g_i, g_j), \quad i = 1, 2, ..., \quad \mu_1 = (\mu_1^1, \mu_2^1, ..) \in l^2.$$

Indeed, if $A^*\mu_1 = u_0$, $\mu_1 \in l^2$ then for $i \in N$,

$$\begin{aligned}
\mu_i^0 &= (u_0, g_i)\\
&= (A^*\mu_1, g_i)\\
&= \left(\sum_{j=1}^{\infty}\mu_j^1g_j, g_i\right)\\
&= \sum_{j=1}^{\infty}\mu_j^1(g_j, g_i).
\end{aligned}$$

Conversely, if these equalities hold for all i=1,2,.. then it is easily seen that $(A^*\mu_1 - u_0, g_i) = 0$, $\forall i \in N$, i.e., $A^*\mu_1 = u_0$.

Concerning the condition in (b) of Theorem 2.7, it is more convenient to use the spectral decomposition of A^*A. As noted above, A^*A is positive and self-adjoint. Let us check that A^*A is a compact operator from $L^2(\Omega)$ to $L^2(\Omega)$. Assume u_n converges to u weakly in $L^2(\Omega)$. We need to show that

$$A^*Au_n \longrightarrow A^*Au \quad (strongly) \ in \ L^2(\Omega). \tag{2.36}$$

In fact, it follows from the weak convergence of (u_n) that

$$\int_{\Omega}u_ng_j \longrightarrow \int_{\Omega}ug_j, \tag{2.37}$$

for each $j \in \mathbf{N}$. Also, $M_1 = \sup\{\|u_n\| : n \in \mathbf{N}\} < \infty$. Let $\epsilon > 0$. Since

$$\sum_{j=1}^{\infty} \|g_j\|^2 < \infty,$$

there exists $n_0 \in \mathbf{N}$ such that

$$\sum_{j=n_0}^{\infty} \|g_j\|^2 < \epsilon.$$

On the other hand, from (2.37), there exists $n_1 \in \mathbf{N}$ such that

$$|(u_n - u, g_j)| = \left| \int_\Omega u_n g_j - \int_\Omega u g_j \right| < \epsilon,$$

for all $n \geq n_1$, all j in $\{1, 2, ..., n_0 - 1\}$. Put $M_2 = \sup\{\|g_j\| : j \in \mathbf{N}\} < \infty$. For $n \geq n_1$, we have

$$
\begin{aligned}
\|A^*Au_n - A^*Au\| &= \left\| \sum_{j=1}^{\infty} (u_n - u, g_j) g_j \right\| \\
&\leq \sum_{j=1}^{\infty} |(u_n - u, g_j)| \, \|g_j\| \\
&\leq \sum_{j=1}^{n_0-1} |(u_n - u, g_j)| \, \|g_j\| + \sum_{j=n_0}^{\infty} \|u_n - u\| \|g_j\|^2 \\
&\leq M_2(n_0 - 1)\epsilon + \|u_n - u\| \sum_{j=n_0}^{\infty} \|g_j\|^2 \\
&\leq \epsilon(M_2(n_0 - 1) + (M_1 + \|u\|)).
\end{aligned}
$$

This shows (2.36) and thus the compactness of A^*A. By the well known spectral theorem for self-adjoint compact operators in Hilbert spaces (cf. Theorem 1.7), A^*A has a sequence of eigenvalues $\{\lambda_n : n \in \mathbf{N}\}$ and a sequence of eigenfunction $\{e_n : n \in \mathbf{N}\}$, which forms an orthonormal basis of $L^2(\Omega)$. Note that since A^*A is positive, $\lambda_n \geq 0$ for all $n \in \mathbf{N}$. We put

$$I = \{n \in \mathbf{N} : \lambda_n = 0\}.$$

Note that $I \neq \emptyset$ if the set spanned by $\{g_j : j \in \mathbf{N}\}$ is not dense in $L^2(\Omega)$. For each $u \in L^2(\Omega)$, u has the expansion with respect to $\{e_n : n \in \mathbf{N}\}$

$$u = \sum_{j=1}^{\infty} u_j g_j,$$

where $u_j = (u, e_j)$. A^*Au is written as a Fourier series

$$A^*Au = \sum_{j=1}^{\infty} \lambda_j u_j e_j = \sum_{j \notin I} \lambda_j (u, e_j) e_j.$$

We show that Cu (C is the square root of A^*A) is given by

$$Cu = \sum_{j=1}^{\infty} \lambda_j^{1/2}(u, e_j)e_j \qquad (2.38)$$

In fact, it is easy to check that the operator C_0 defined by the right hand side of the equation in (2.38) is well defined, linear, self-adjoint, and satisfies

$$C_0^2 u = C_0(C_0 u) = \sum_{j=1}^{\infty} \lambda_j u_j e_j = A^*A u.$$

From the uniqueness of the square root operator, $C = C_0$ and we have (2.38). From that equation, we immediately have $Ce_j = \lambda_j^{1/2} e_j$ for all $j \in \mathbf{N}$.

The condition $u_0 = Cu_1$ in Theorem 2.7 (b) is now equivalent to

$$(u_0, e_j) = (Cu_1, e_j) = (u_1, Ce_j) = \lambda_j^{1/2}(u_1, e_j), \quad \text{for all } j \in \mathbf{N}.$$

This is equivalent to

$$(u_0, e_j) = 0 \quad \text{for all } j \in \mathbf{N} \qquad (2.39)$$

and

$$(u_1, e_j) = \lambda_j^{-1/2}(u_0, e_j), \quad \text{for all } j \notin I.$$

By the Riesz-Fisher theorem (see Theorem 1.6), the sequence $((u_1, e_j))_{j \in \mathbf{N}}$ determines a function in $L^2(\Omega)$ (with respect to the orthonormal basis (e_j)) if and only if

$$\sum_{j=1}^{\infty} (u_1, e_j)^2 < \infty.$$

Therefore, $u_0 = Cu_1$ for some $u_1 \in L^2(\Omega)$ if and only if (2.39) holds and

$$\sum_{j \in \mathbf{N} \setminus I} \lambda_j^{-1}(u_0, e_j)^2 < \infty.$$

Note that (2.39) is equivalent to $u_0 \perp Ker(A^*A)$.

2.2.2 Case 2: exact solutions in $L^{\alpha^*}(\Omega)$, $1 < \alpha^* < \infty$

So far, we have considered moment problems in $L^2(\Omega)$ (cf. Eq. (2.1)). It turns out that moment problems can in a natural way be formulated in $L^\alpha(\Omega)$ for $1 < \alpha < \infty$. The problem is then to find a function u in $L^{\alpha^*}(\Omega)$ such that

$$\int_\Omega u(y)g_j(y)dy = \mu_j, \quad j = 1, 2, \dots. \qquad (MP)$$

Here α^* is the conjugate exponent of α, Ω is as before a bounded domain in R^d, $g_1, g_2, \dots$ is a given sequence of functions in $L^\alpha(\Omega)$ and $\mu_1, \mu_2, \dots$ is a sequence of real numbers. The problem is most conveniently formulated as a linear equation in a reflexive Banach space.

The problem in a Banach space setting.

We consider the equation (2.30) where $(X, \|.\|)$, $(Y, \|.\|)_Y$ are reflexive Banach spaces, and $A: X \to Y$ is linear and continuous.

By known properties of reflexive Banach spaces (see e.g. Ch. 2, [Li]), there exists a norm $\|.\|_0$ on X (resp. $\|.\|_1$ on Y) which is equivalent to $\|.\|$ (resp. $\|.\|_Y$), such that $(X, \|.\|_0)$ (resp. $(Y, \|.\|_1)$) is strictly convex.

We have, in this case, the following results (see e.g. Sec. 2.2, Ch. 2, [Li])

(i) $(X, \|.\|_0)$ (resp. $(Y, \|.\|_1)$) and its dual $(X^*, \|.\|_0^*)$ (resp. $(Y^*, \|.\|_1^*)$) are strictly convex .

(ii) If $x_n \rightharpoonup x$ in X-weak (resp. $y_n \rightharpoonup y$ in Y-weak) and $\|x_n\|_0 \to \|x\|_0$ (resp. $\|y_n\|_1 \to \|y\|_1$) then $x_n \to x$ in X-strong (resp. $y_n \to y$ in Y-strong).

(iii) For each $x \in X$ (resp. $y \in Y$) there exists a unique $J(x) \in X^*$ (resp. $L(y) \in Y^*$) such that

$$\|J(x)\|_0^* = \|x\|_0 \quad and \quad < J(x), x >= \|x\|_0^2$$

(resp.

$$\|L(y)\|_1^* = \|y\|_1 \quad and \quad < L(y), y >= \|y\|_1^2).$$

Here, we use $< ., . >$ to denote both the pairings between X, X^* and Y, Y^*.

The mapping $J: X \to X^*$ (resp. $L: Y \to Y^*$) is called the duality mapping corresponding to $\|.\|_0$ (resp. $\|.\|_1$). We know that J (resp. L) is a homeomorphism of X onto X^* (resp. of Y onto Y^*).

For $\beta > 0$, consider the following problem of finding $u \in X$ such that

$$\beta < J(u), v > + < L(Au), Av >=< L(\mu), Av >, \qquad \forall v \in X. \tag{2.40}$$

Proposition 2.8. *For each $\beta > 0$ fixed, (2.40) has a unique solution $u = u^\beta(\mu)$ depending continuously on μ.*

Proof. From (i)-(iii), we see that $\beta J + A^* L A$ is coercive and strictly monotone from X to X^*. The existence and uniqueness of a solution of (2.40) thus follow from well-known results concerning variational inequalities (see e.g., Theorem 8.2, 8.3, [Li]). We now prove the continuous dependence of $u^\beta(\mu)$ on μ. Suppose by contradiction that there is a sequence $(\mu^n) \subset Y$ such that $\mu^n \to \mu$ in Y and that

$$\|u^\beta(\mu^n) - u^\beta(\mu)\| \geq \epsilon_0 > 0, \qquad \forall n. \tag{2.41}$$

Put $u_n = u^\beta(\mu^n)$ and let $\mu = \mu^n$, $v = u_n$ in (2.40). We have

$$\beta\|u_n\|_0^2 + \|Au_n\|_1^2 = < L(\mu^n), Au_n >$$
$$\leq \|\mu^n\|_1 \|u_n\|_0 \|A\|$$

($\|A\|$ is the norm of $A : (X, \|.\|_0) \to (Y, \|.\|_1)$). Since (μ^n) is bounded, so is (u_n). Hence we can take a subsequence $(u_{n_k}) \subset (u_n)$ such that

$$u_{n_k} \rightharpoonup u_0 \quad in \ X - weak. \tag{2.42}$$

Letting $\mu = \mu^{n_k}$, $v = u_{n_k} - u_0$ in (2.40), we get

$$\beta < J(u_{n_k}), u_{n_k} - u_0 > + < L(Au_{n_k}) - L(Au_0), Au_{n_k} - Au_0 >$$

$$= < L(\mu^{n_k}) - L(Au_0), A(u_{n_k} - u_0) > . \tag{2.43}$$

But from (2.42), $Au_{n_k} \rightharpoonup Au_0$ in Y-weak, and thus, from $L(\mu^{n_k}) \to L(\mu)$ in Y-strong, one has

$$< L(\mu^{n_k}) - L(Au_0), A(u_{n_k} - u_0) > \longrightarrow 0 \quad as \ k \to \infty.$$

Letting $k \to \infty$ in (2.43), we obtain

$$\limsup < J(u_{n_k}) - J(u_0), u_{n_k} - u_0 > =$$
$$= \limsup < J(u_{n_k}), u_{n_k} - u_0 > \le 0. \tag{2.44}$$

But, it is easy to check that

$$< J(u_{n_k}) - J(u_0), u_{n_k} - u_0 > = (\|u_{n_k}\|_0 - \|u_0\|_0)^2$$
$$+ (\|J(u_{n_k})\|_0^* \|u_0\|_0 - < J(u_{n_k}), u_0 >)$$
$$+ (\|J(u_0)\|_0^* \|u_{n_k}\|_0 - < J(u_0), u_{n_k} >)$$
$$\ge (\|u_{n_k}\|_0 - \|u_0\|_0)^2.$$

Hence

$$\|u_{n_k}\|_0 \to \|u_0\|_0, \quad k \to \infty.$$

This, together with (ii) and (2.42) gives

$$u_{n_k} \to u_0 \quad in \ X - strong. \tag{2.45}$$

Now, letting $\mu = \mu^{n_k}$ in (2.40), and letting $k \to \infty$, taking account of this limit, we have

$$\beta < J(u_0), v > + < L(Au_0), Av > = < L(\mu), Av >, \quad \forall v \in X,$$

i.e., u_0 is a solution of (2.40).

By the uniqueness of the solution of (2.40), we must have $u_0 = u^\beta(\mu)$. This and (2.45) contradict (2.41), which completes the proof of Proposition 2.8.

An error estimate

Since A may not be one-to-one, the set of solutions of (2.30) may contain more than one element. But this set is closed and convex in X. Hence, by the strict convexity of $\|.\|_0$, it can be seen that if (2.30) is solvable then it has a unique solution which minimizes $\|.\|_0$. In the following theorem, we show that this solution can be approximated by those of (2.40).

Let $\mu^0 \in Y$ be such that (2.30) with $\mu = \mu^0$ has a solution. Let u_0 be its solution with minimal $\|.\|_0$-norm. Let φ be the modulus of continuity at μ^0 of L (which, we recall, is the duality mapping as in (2.40))

$$\varphi(t) = \sup \left\{ \|L(\mu) - L(\mu^0)\|_1^* : \ \|\mu - \mu^0\| \le t \right\}, \quad t > 0.$$

The function φ depends only on L and μ^0. We have $\varphi(t) > 0$, $\forall t > 0$, and $\varphi(t) \to 0$ as $t \to 0$.

Theorem 2.9. *Let $\beta(\epsilon) = [\varphi(\epsilon)]^{1/2}$ ($\epsilon > 0$). Then, for each $\eta > 0$, there exists an $\epsilon > 0$ such that for all $\mu \in Y$ satisfying $\|\mu - \mu^0\|_Y \le \epsilon$, we have*

$$\|u^{\beta(\epsilon)}(\mu) - u_0\| \le \eta.$$

Proof. Suppose by contradiction that there exist an $\eta_0 > 0$ and a sequence $(\mu^n) \subset Y$ such that

$$\|\mu^n - \mu^0\|_Y < 1/n \text{ and } \|u_n - u_0\| > \eta_0, \quad \forall n.$$

Here $u_n = u^{\beta(1/n)}(\mu^n)$. Letting $\mu = \mu^n$, $v = u_n - u_0$ in (2.40), one has

$$\beta(1/n) < J(u_n), u_n - u_0 > + < L(Au_n), Au_n - Au_0 >$$

$$= < L(\mu^n), A(u_n - u_0) > . \tag{2.46}$$

Since

$$< L(Au_0), A(u_n - u_0) > = < L(\mu_0), A(u_n - u_0) >,$$

we get

$$(\varphi(1/n))^{1/2} < J(u_n), u_n - u_0 > + < L(Au_n) - L(Au_0), Au_n - Au_0 >$$

$$= < L(\mu_n) - L(\mu^0), A(u_n - u_0) >$$
$$\le \|L(\mu_n) - L(\mu^0)\|_1^* \|A\| \|u_n - u_0\|_0$$
$$\le \frac{1}{2}(\varphi(1/n))^{-1}(\|L(\mu_n) - L(\mu^0)\|_1^*)^2 +$$
$$+ \frac{1}{2}\varphi(1/n)\|A\|^2(\|u_n\|_0 + \|u_0\|_0)^2$$
$$\le \frac{1}{2}\varphi(1/n) + \varphi(1/n)\|A\|^2(\|u_n\|_0^2 + \|u_0\|_0^2)$$

(remark that

$$\|L(\mu^n) - L(\mu^0)\|_1^* \le \varphi(\|\mu^n - \mu^0\|_Y) \le \varphi(1/n)).$$

Hence, by the monotony of L, we have

$$< J(u_n), u_n - u_0 > \leq \frac{1}{2}(\varphi(1/n))^{1/2} + (\varphi(1/n))^{1/2}\|A\|^2(\|u_n\|_0^2 + \|u_0\|_0^2). \quad (2.47)$$

For n sufficiently large (so that $(\varphi(1/n))^{1/2}\|A\|^2 \leq 1/2$), we have

$$\begin{aligned}
\|u_n\|_0^2 &=< J(u_n), u_n > \\
&\leq \frac{1}{2}(\varphi(1/n))^{1/2} + \frac{1}{2}(\|u_n\|_0^2 + \|u_0\|_0^2) + < J(u_n), u_0 > \\
&\leq \frac{1}{2}(\varphi(1/n))^{1/2} + \frac{1}{2}(\|u_n\|_0^2 + \|u_0\|_0^2) + \|u_n\|_0\|u_0\|_0 \\
&\leq \frac{1}{2}(\varphi(1/n))^{1/2} + \frac{3}{4}\|u_n\|_0^2 + \frac{3}{2}\|u_0\|_0^2.
\end{aligned}$$

This means that (u_n) is bounded in X. Taking account of (2.47), we get

$$\limsup < J(u_n), u_n - u_0 > \leq 0. \quad (2.48)$$

But

$$\begin{aligned}
< J(u_n), u_n - u_0 > &= \|u_n\|_0^2 - < J(u_n), u_0 > \\
&\geq \|u_n\|_0(\|u_n\|_0 - \|u_0\|_0),
\end{aligned}$$

i.e.

$$\|u_n\|_0 - \|u_0\|_0 \geq \|u_0\|_0^{-1} < J(u_n), u_n - u_0 > .$$

Hence

$$\limsup(\|u_n\|_0 - \|u_0\|_0) \geq 0,$$

i.e.,

$$\limsup \|u_n\|_0 \geq \|u_0\|_0. \quad (2.49)$$

On the other hand, by the boundedness of (u_n), we can choose a subsequence $(u_{n_k}) \subset (u_n)$ such that

$$u_{n_k} \rightharpoonup u^* \quad in \ X - weak. \quad (2.50)$$

It follows that

$$Au_{n_k} \rightharpoonup Au^* \quad in \ Y - weak \quad (2.51)$$

and that

$$\|u^*\|_0 \leq \liminf \|u_{n_k}\|_0. \quad (2.52)$$

Now, letting $\mu = \mu^{n_k}$, $v = u_{n_k} - u^*$ in (2.40) (with $\beta = \beta(1/n_k)$) we have, for all k,

$$(\varphi(1/n_k))^{1/2} < J(u_{n_k}), u_{n_k} - u^* > + < L(Au_{n_k}), Au_{n_k} - Au^* >$$

$$=< L(\mu^{n_k}), A(u_{n_k} - u^*) > .$$

Since (u_{n_k}) is bounded, $\beta(1/n_k) \to 0$ and $L(\mu^{n_k}) \to L(\mu^0)$ (strong) as $k \to \infty$, we have

$$(\varphi(1/n_k))^{1/2} < J(u_{n_k}), u_{n_k} - u^* > \to 0 \quad as \ k \to \infty$$

and

$$< L(\mu^{n_k}), A(u_{n_k} - u^*) > \to 0 \quad as \ k \to \infty.$$

Now, letting $k \to \infty$ in the latter relation, we obtain

$$\lim < L(Au_{n_k}), A(u_{n_k} - u^*) >= 0.$$

Arguing as in the proof of (2.45), we have from (2.51),

$$A(u_{n_k}) \to Au^* \quad in \ Y - strong. \tag{2.53}$$

Consequently,

$$L(Au_{n_k}) \to L(Au^*) \quad in \ Y - strong.$$

Now, letting $\mu = \mu^{n_k}$, $\beta = \beta(1/n_k)$ in (2.40) and then letting $k \to \infty$, one gets

$$< L(Au^*), Av >=< L(\mu^0), Av >=< L(Au_0), Av >, \quad \forall v \in X.$$

With $v = u^* - u_0$, this gives

$$< L(Au^*) - L(Au_0), Au^* - Au_0 >= 0.$$

Since L is strictly monotone, this implies

$$Au^* = Au_0 = \mu^0,$$

i.e., u^* is a solution of (2.30).

On the other hand, we have, by (2.49) and (2.52),

$$\|u^*\|_0 \leq \|u_0\|_0.$$

By the definition of u, one must have $u^* = u_0$. Using arguments similar to those leading to (2.53), we obtain from this, (2.48), (2.50), that

$$u_{n_k} \to u_0 \quad in \ X - strong.$$

But this contradicts the choice of (u_n) and completes the proof of Theorem 2.9.

Remark 2.1. When Y is a Hilbert space, $L = I_Y$ is the identity mapping of Y. Hence $\varphi(t) = t$ $(t > 0)$ and in Theorem 2.9, $\beta(\epsilon) = \sqrt{\epsilon}$ $(\epsilon > 0)$.

The moment problem in an $L^{\alpha^*}(\Omega)$-setting.

We now consider the moment problem in $L^\alpha(\Omega)$, $1 < \alpha < \infty$, and state it as follows.

Let $g_1, g_2, \ldots$ be a linear independent sequence of functions in $L^\alpha(\Omega)$. Let $\mu = (\mu_1, \mu_2, \ldots)$ be a sequence of real numbers. We are to find a function $u \in L^{\alpha^*}(\Omega)$ (α^* is the conjugate exponent of α) such that

$$\int_\Omega u g_j = \mu_j, \quad j \in \mathbf{N}. \tag{2.54}$$

By dividing both sides of (2.54) by appropriate constants (as in the beginning of Sec. 2.1.), we can assume in addition that

$$\sum_{j=1}^{\infty} \|g_j\|^2_{L^{\alpha}(\Omega)} < \infty.$$

Under these conditions, one can check that the operator A defined by (2.29) is linear and continuous from $L^{\alpha^*}(\Omega)$ to l^2. In this case, $X = L^{\alpha^*}(\Omega)$ is a uniformly convex Banach space with the usual norm $\|.\|_{L^{\alpha^*}(\Omega)}$, i.e.

$$\|.\|_0 = \|.\| = \|.\|_{L^{\alpha^*}(\Omega)}.$$

The duality mapping

$$J: \ L^{\alpha^*}(\Omega) \to L^{\alpha}(\Omega) \ (= [L^{\alpha^*}(\Omega)]^*)$$

is given by

$$J(v) = \|v\|^{(\alpha-2)/(\alpha-1)}_{L^{\alpha^*}(\Omega)} |v|^{1/(\alpha-1)} \mathrm{sgn} v, \quad \text{for} \ \ v \neq 0, \ \ J(0) = 0, \ \ v \in L^{\alpha^*}(\Omega).$$

On the other hand, $Y = l^2$ is a Hilbert space, and as remarked above, $L(y) = y$, $\forall y \in Y$ and $\beta(\epsilon) = \sqrt{\epsilon}$ ($\epsilon > 0$). Theorem 2.9 gives us a stabilized approximation of the solution u_0 of (2.54) which minimizes the usual norm $\|.\|_{L^{\alpha^*}(\Omega)}$.

2.2.3 Case 3: exact solutions in $H^1(\Omega)$

In this subsection, we regularize (2.1) by variational equations similar to (2.31). We assume here that $< g_1, g_2, .. >$ is dense in $L^2(\Omega)$. Hence A (defined by (2.29)) is one-to-one from $L^2(\Omega)$ to l^2. We denote by $(.,.)_1$ and $\|.\|_1$ the usual inner product and norm on $H^1(\Omega)$.

For $\beta > 0$, consider the following problem of finding $u \in H^1(\Omega)$ such that

$$\beta(u,v)_1 + (Au, Av)_{l^2} = (\mu, Av)_{l^2}, \qquad \forall v \in H^1(\Omega). \tag{2.55}$$

The following stability result is similar to Proposition 2.6 and Proposition 2.8, and its proof is omitted.

Proposition 2.10. *For each $\beta > 0$, (2.55) has a unique solution $u = u^{\beta}(\mu)$ which depends continuously (with respect to $\|.\|_1$) on $\mu \in l^2$.*

Now, since A is injective and since the embedding

$$H^1(\Omega) \hookrightarrow L^2(\Omega)$$

is compact, we claim that the mapping

$$\varphi: \ [0,\infty) \longrightarrow [0,\infty)$$

defined by

$$\varphi(t) = t + \sup_{x \in B(t)} \|x\|, \qquad t \geq 0, \tag{2.56}$$

$$B(t) = \{x \in H^1(\Omega) : \ \|x\|_1 \leq 1 \ \text{and} \ \|Ax\|_{l^2} \leq t\}, \qquad t \geq 0.$$

is increasing, and continuous at 0, and $\varphi(0) = 0$. In fact, for $0 < t_1 < t_2$, if $x \in B(t_1)$ then

$$\|x\|_1 \leq 1 \ and \ \|Ax\|_{l^2} \leq t_1 < t_2,$$

i.e., $x \in B(t_2)$. Hence $B(t_1) \subset B(t_2)$, which gives

$$\sup_{x \in B(t_1)} \|x\| \leq \sup_{x \in B(t_2)} \|x\|.$$

From the latter inequality, we get

$$\varphi(t_1) = t_1 + \sup_{x \in B(t_1)} \|x\| \leq t_2 + \sup_{x \in B(t_2)} \|x\| = \varphi(t_2),$$

i.e., φ is increasing. We now prove that $\varphi(0) = 0$. If $x \in B(0)$, one has by the definition of $B(t)$ that $Ax = 0$. Since A is injective, the latter equality implies $x = 0$. Hence, $B(0) = \{0\}$ and $\varphi(0) = 0 + \sup_{x \in B(0)} \|x\| = 0$. Finally, we prove that φ is continuous at $t = 0$. Suppose by contradiction that φ is not continuous at $t = 0$. Then, there exists a sequence (t_n) of nonnegative numbers and an $\epsilon_0 > 0$ such that

$$t_n \downarrow 0 \ and \ \varphi(t_n) \geq \epsilon_0 > 0. \tag{2.57}$$

By the definition of φ, for each n we can find an $x_n \in B(t_n)$ such that

$$\varphi(t_n) \geq t_n + \|x_n\| + \frac{1}{n}. \tag{2.58}$$

Letting $n \to \infty$ in (2.57), (2.58), we get

$$\liminf_{n \to \infty} \|x_n\| \geq \epsilon_0. \tag{2.59}$$

On the other hand, since the embedding

$$H^1(\Omega) \hookrightarrow L^2(\Omega)$$

is compact, we can find a subsequence $(x_{n_k}) \subset (x_n)$ and an $x_0 \in L^2(\Omega)$ such that $x_{n_k} \to x_0$ in $L^2(\Omega)$. Noting that

$$\|Ax_{n_k}\|_{l^2} \leq \frac{1}{n_k}$$

we get after letting $k \to \infty$ in the latter inequality that $Ax_0 = 0$, which gives $x_0 = 0$. This contradicts (2.59). Hence φ is continuous at $t = 0$.

The following theorem gives a regularization of (2.1) by the solution of (2.55)

Theorem 2.11. *Suppose the solution u_0 of (2.1) corresponding to $\mu = \mu^0$ is in $H^1(\Omega)$. Then, by choosing $\beta = \beta(\epsilon) = \epsilon$, $0 < \epsilon < 1$, we have, for all μ satisfying*

$$\|\mu - \mu^0\|_{l^2} \leq \epsilon,$$

the estimate

$$\|u^{\beta(\epsilon)}(\mu) - u_0\| \leq (1 + \|u_0\|_1)\varphi(\epsilon^{1/2}).$$

Proof. Since $H^1(\Omega)$ is dense in $L^2(\Omega)$, (2.1) (with $u = u_0$, $\mu = \mu^0$) implies

$$(Au_0, Av)_{l^2} = (\mu, Av)_{l^2}, \qquad \forall v \in H^1(\Omega).$$

Subtracting this equality from (2.55), and letting

$$u = u^{\beta(\epsilon)}(\mu), \; v = u - u_0, \; \beta = \beta(\epsilon) = \epsilon,$$

we get

$$\begin{aligned}
\epsilon\|u - u_0\|_1^2 + \|A(u - u_0)\|_{l^2}^2 &= \\
&= (\mu - \mu^0, A(u - u_0))_{l^2} + \epsilon(u_0, u - u_0)_1 \\
&\leq \|\mu - \mu^0\|_{l^2}\|A(u - u_0)\|_{l^2} + \epsilon\|u_0\|_1\|u - u_0\|_1 \\
&\leq \frac{1}{4}\|\mu - \mu^0\|_{l^2}^2 + \|A(u - u_0)\|_{l^2}^2 + \\
&\quad + \frac{1}{2}\epsilon(\|u_0\|_1^2 + \|u - u_0\|_1^2).
\end{aligned} \tag{2.60}$$

Hence

$$\begin{aligned}
\|u - u_0\|_1^2 &\leq \frac{1}{2}\epsilon + \|u_0\|_1^2 \\
&\leq 1 + \|u_0\|_1^2 \\
&\leq (1 + \|u_0\|_1)^2,
\end{aligned}$$

i.e.,

$$\|u - u_0\|_1 \leq 1 + \|u_0\|_1. \tag{2.61}$$

On the other hand, one can check from (2.56) that φ is strictly increasing from $[0, \infty)$ onto $[0, \infty)$, and that

$$\|x\| \leq \varphi(\|Ax\|_{l^2}),$$

i.e.,

$$\|Ax\|_{l^2} \geq \varphi^{-1}(\|x\|), \quad \forall x \in H^1(\Omega), \|x\|_1 \leq 1.$$

By (2.61), we have $\|x\|_1 \leq 1$ with

$$x = (1 + \|u_0\|_1)^{-1}(u - u_0)$$

In particular,

$$(1 + \|u_0\|_1)^{-1}\|A(u - u_0)\|_{l^2} \geq \varphi^{-1}\left(\frac{\|u - u_0\|_1)}{1 + \|u_0\|_1}\right). \tag{2.62}$$

On the other hand, it can be seen from (2.60) that

$$\begin{aligned}
\epsilon\|u - u_0\|_1^2 + \|A(u - u_0)\|_{l^2}^2 &= \\
&\leq \frac{1}{2}\|\mu - \mu^0\|_{l^2}^2 + \frac{1}{2}\|A(u - u_0)\|_{l^2}^2 + \\
&\quad + \frac{1}{4}\epsilon\|u_0\|_1^2 + \epsilon\|u - u_0\|_1^2.
\end{aligned}$$

Hence

$$\|A(u - u_0)\|_{l^2}^2 \leq \epsilon^2 + \frac{1}{2}\epsilon\|u_0\|_1^2$$
$$\leq \epsilon(1 + \|u_0\|_1^2),$$

i.e.,

$$\|A(u - u_0)\|_{l^2} \leq \epsilon^{1/2}(1 + \|u_0\|_1).$$

This and (2.62) imply

$$\varphi^{-1}\left(\frac{\|u - u_0\|_1)}{1 + \|u_0\|_1}\right) \leq \epsilon^{1/2},$$

that is,

$$\|u - u_0\|_1 \leq (1 + \|u_0\|_1)\varphi(\epsilon^{1/2}).$$

Theorem 2.11 is proved.

Remark 2.2.

(i) We can replace $H^1(\Omega)$ by $H^m(\Omega)$ in Theorem 2.11 without any notable change. By choosing $m \geq 1$ such that $H^m(\Omega) \hookrightarrow L^{\alpha^*}(\Omega)$ is compact, we can apply the above arguments to the moment problem in $L^\alpha(\Omega)$.

(ii) Since they are coercive, the variational equations (2.31), (2.40), (2.55) can be solved by the usual Galerkin method. Detailed discussions can be found, e.g., in [Ci].

2.3 Notes and remarks

At the beginning of the chapter, we gave two examples of moment problems, namely, the Hausdorff moment problem and the trigonometric problem. Following are a few more examples. Consider the problem of finding u in $L^2(\mathbf{R}_+)$ satisfying the sequence of equations

$$\int_0^\infty u(x)e^{-nx}dx = \mu_n, \qquad n \in \mathbf{N}. \tag{2.63}$$

where (μ_n) is a given real sequence. Then the mapping A associated with this moment problem

$$(Au)(n) = \int_0^\infty u(x)e^{-nx}dx, \qquad n \in \mathbf{N}$$

takes $L^2(\mathbf{R}_+)$ into l^2. Indeed, by a result of Hilbert and Hardy [DS], p.533, the mapping A_0 defined by

$$(A_0u)(p) = \int_0^\infty u(x)e^{-px}dx, \qquad p > 0 \tag{2.64}$$

takes $L^2(\mathbf{R}_+)$ into itself, from which it immediately follows that A takes $L^2(\mathbf{R}_+)$ into l^2, as claimed.

Now consider the Stieltjes transform

$$(A\phi)(x) = \int_0^\infty \frac{\phi(t)dt}{x+t} \qquad x > 0. \tag{2.65}$$

The function $A\phi(x)$ can be extended to a complex analytic function on a strip about the positive real axis of the complex plane. Hence if (x_n) is any bounded sequence of positive numbers such that $x_i \neq x_j$ for $i \neq j$, then the moment problem

$$\int_0^\infty \frac{\phi(t)dt}{x_n+t} = \mu_n, \qquad n \in \mathbf{N}, \tag{2.66}$$

admits at most one solution in $L^2(\mathbf{R}_+)$.

Now let

$$A : E \longrightarrow F$$

be a continuous injective linear map of a Banach space E into a Banach space F (both E and F are infinite dimensional). If A is compact then $A^{-1} : A(F) \longrightarrow E$ is unbounded. As we will see below, there are infinitely many examples of noncompact maps A such that A^{-1} is unbounded.

Our first example of a noncompact operator A such that A^{-1} is unbounded is the Fourier transform $\mathcal{F}$ of $L^1(\mathbf{R})$ into $C_0(\mathbf{R})$, the Banach space of functions continuous on $\mathbf{R}$ vanishing at infinity endowed with the sup-norm topology

$$\mathcal{F} : L^1(\mathbf{R}) \longrightarrow C_0(\mathbf{R})$$

$$\mathcal{F}f(t) = \int_{-\infty}^\infty f(x)e^{-itx}dx.$$

We first have that $\mathcal{F}^{-1}$ is unbounded. Indeed the range of $\mathcal{F}$ is a dense proper subspace of $C_0(\mathbf{R})$ and thus by a theorem of Banach $\mathcal{F}^{-1}$ is not continuous. We show next that $\mathcal{F}$ is noncompact. Indeed, consider the sequence $f_n(x) = \chi_{[n,n+1]}(x)$ where $\chi_{[n,n+1]}$ is the characteristic function of $[n, n+1]$. We have $\|f_n\|_{L^1(\mathbf{R})} = 1$. We claim that the closure of $\{\mathcal{F}f_n\}$ is noncompact. In fact writing $\hat{f}_n$ for $\mathcal{F}f_n$, we have

$$\hat{f}_n = \int_n^{n+1} e^{-ixt}dx = e^{-int}\left(\frac{1-e^{-it}}{it}\right). \tag{2.67}$$

Suppose by contradiction that there exists a subsequence (f_{n_k}) such that $(\hat{f}_{n_k})$ converge to a g in $C_0(\mathbf{R})$. Then for each interval $(a,b) \subset \mathbf{R}$, we have that for all ϕ in $L^2(a,b)$,

$$\int_a^b e^{-int}\left(\frac{1-e^{-it}}{it}\right)\phi(t)dt \longrightarrow \int_a^b g\phi \, dt \qquad for \ n \to \infty$$

On the other hand, we have

$$\int_a^b e^{-int}\left(\frac{1-e^{-it}}{it}\right)\phi(t)dt \longrightarrow 0 \qquad for \ n \to \infty$$

It follows that

$$\int_a^b g\phi dt = 0 \qquad \text{for all } \phi \in L^2(a,b).$$

Thus $g = 0$ on (a,b). Since (a,b) is arbitrary, we deduce that $g \equiv 0$ on $\mathbf{R}$, and hence that

$$\hat{f}_{n_k} \longrightarrow 0 \qquad \text{in } C_0(\mathbf{R}) \ \text{for } k \to \infty.$$

Therefore

$$\|\hat{f}_{n_k}\|_\infty \longrightarrow 0 \qquad \text{for } k \to \infty. \tag{2.68}$$

By (2.67)

$$\|\hat{f}_{n_k}\|_\infty = \sup_{t\in\mathbf{R}} \left| \frac{1 - e^{-it}}{it} \right| = const > 0$$

which contradicts (2.68). This contradiction shows that $(\hat{f}_n)$ is not compact. We conclude that $\mathcal{F}$ is a noncompact map.

Concerning the operator A of the Hausdorff moment problem taking $L^2(0,1)$ into l^2, it is known that A^{-1} is unbounded. B. Hofmann [H] has raised the question whether A is or is not compact. The answer which is negative is given by A. Neubauer to whom we owe the counterexample $x_n(t) = \sqrt{n}\ t^n$.

For completeness, we show that A^{-1} is unbounded. In fact, consider the operator $A: L^2(0,1) \to l^2$ defined by

$$Au = (\mu_0, \mu_1, ...)$$

with

$$\mu_k = \int_0^1 x^k u(x) dx.$$

We claim that A^{-1} is unbounded. In fact, let $u_p(x) = x^{-1/2-\ln x/p}$, $p = 1, 2, \ldots$. Then

$$\mu_k = \int_0^1 x^{k-1/2-\ln x/p} dx$$

$$= \int_0^\infty e^{-t(k+1/2)-t^2/p} dt \leq \frac{1}{k+1/2}.$$

Hence,

$$\|Au_p\|_{l^2} \leq \sum_{k=0}^\infty \left(k + \frac{1}{2}\right)^{-2} < \infty \quad \forall p > 0.$$

But we have after some computations

$$\|u_p\|_{L^2}^2 = \sqrt{p/2} \int_0^\infty e^{-y^2} dy \to \infty \qquad \text{for } p \to \infty.$$

It follows that A^{-1} is unbounded (cf. Inglese [In]).

On the other hand, A is not compact. In fact, let $x_n(t) = \sqrt{n}t^n$, $n = 1, 2, \ldots$. Then all $x_n \in L^2(0,1)$. We shall prove that $x_n \rightharpoonup 0$ as $n \to \infty$ and

$$\|Ax_n\|_{l^2}^2 \longrightarrow \int_0^\infty \frac{dx}{1+x^2} > 0.$$

It will then follow that A is not compact.

We first prove that $x_n \rightharpoonup 0$ in $L^2(0,1)$ as $n \to \infty$. Indeed, let $\varphi \in L^2(0,1)$ and let $0 < \delta_n < 1$, $\delta_n \to 0$ as $n \to \infty$, be a sequence of numbers to be specified later and let

$$\varphi_n(t) = \varphi(t) \qquad \text{for } 1 - \delta_n < t < 1,$$
$$= 0 \qquad \text{for } 0 < t < \delta_n$$

Then

$$\left| \int_0^1 x_n(t)\varphi(t)dt \right| = \left| \left(\int_0^{1-\delta_n} + \int_{1-\delta_n}^1 \right) \sqrt{n}t^n \varphi(t)dt \right|$$

$$\leq \left(\int_0^{1-\delta_n} nt^{2n}dt \right)^{1/2} \|\varphi\|_{L^2} + \left(\int_{1-\delta_n}^1 nt^{2n}dt \right)^{1/2} \|\varphi_n\|_{L^2}$$

$$\leq (1 - \delta_n)^{n+1/2}\|\varphi\|_{L^2} + \|\varphi_n\|_{L^2}.$$

Now, for the choice $\delta_n = 1/\sqrt{n}$, one has

$$\|\varphi_n\|_{L^2}^2 = \int_{1-\delta_n}^1 |\varphi(t)|^2 dt \to 0 \quad \text{as } n \to \infty$$

and

$$(1 - \delta_n)^{n+1/2} \to 0 \quad \text{as } n \to \infty.$$

It follows that $x_n \rightharpoonup 0$ in $L^2(0,1)$.

On the other hand,

$$\mu_n = (Ax_n)_k = \int_0^1 t^k \sqrt{n}t^n dt = \frac{\sqrt{n}}{n+k+1}.$$

Hence,

$$\|Ax_n\|_{l^2}^2 = n \sum_{k=0}^\infty \frac{1}{(n+k+1)^2}$$

$$= \sum_{k=0}^\infty \frac{1/n}{(1 + (k+1)/n)^2}$$

$$\to \int_0^\infty \frac{dx}{1+x^2} > 0 \quad \text{as } n \to \infty.$$

From the above argument, it follows that A is not compact as claimed.

We conclude with an example of a moment problem of the form (0.2) of the Introduction, corresponding to $\Omega = (0, 2\pi)$, $g_n(t) = e^{int}$, i.e.,

$$\int_0^{2\pi} e^{int} d\sigma(t) = \mu_n, \qquad n = 0, 1, 2, ..., \bar{\mu}_n = c_{-n}, \tag{2.69}$$

where the unknown function $\sigma(t)$ is a nondecreasing function on $[0, 2\pi]$. This is a trigonometric moment problem, a special case of which was given in (2.3). Trigonometric moment problems are discussed in, e.g., [AGl], [RN], [Kr]. We have

Proposition 2.12. *(i) Suppose there exists a nondecreasing function $\sigma(t)$ on $[0, 2\pi]$ satisfying (2.69). Then the nonnegativity of the trigonometric sum*

$$\sum_{n=-k}^{n=k} \xi_n e^{int}, \qquad k = 0, 1, 2, \ldots, \text{ for all } t \in [0, 2\pi],$$

implies the inequality

$$\sum_{n=-k}^{n=k} \xi_n \mu_n \geq 0.$$

(ii) Suppose that for any real number x, the expressions

$$\sum_{n=-k}^{n=k} \left(1 - \frac{|n|}{k}\right) e^{-inx} \mu_n, \qquad n = 1, 2, ...,$$

are nonnegative. Then there exists a nondecreasing function $\sigma(t)$ on $[0, 2\pi]$ satisfying (2.69).

A proof of the preceding proposition is given in [AGl] where the problem of the number of solutions of (2.69) is also discussed.

3 Backus-Gilbert regularization of a moment problem

3.1 Introduction

In Chapter 2, we presented two methods of regularization for the moment problem, namely, the method of truncated expansion in $L^2(\Omega)$ and the method of regularization by coercive variational equations (the Tikhonov method) in $L^p(\Omega), 1 < p < \infty$.

In the method of truncated expansion of Chapter 2, we used an orthonormalization of the system (g_n). In the present chapter, we also use truncated expansion, however, the approximate solutions, instead of being built from an orthonormalized system, are constructed as combinations of some predetermined basis functions (called the *Backus-Gilbert basis functions*). The natural framework for this method is an $L^p(\Omega)$ space for $1 < p < \infty$. In Chapter 4 (Notes and remarks) we shall apply the Backus-Gilbert method to regularize the Hausdorff moment problem in an L^p-setting, $1 < p < \infty$. Thus we shall consider the problem

(MP) *Find a function u in $L^p(\Omega)$ such that*

$$\int_\Omega u(y)g_j(y)dy = \mu_j, \qquad j = 1, 2, \ldots,$$

where Ω is a bounded domain in $\mathbf{R}^d$, q is the conjugate exponent of p, $1 < p < \infty$, $g_1, g_2, \ldots$ is a given sequence of functions in $L^q(\Omega)$ and $(\mu_1, \mu_2, \ldots)$ is a sequence of real numbers.

We shall briefly go over the motivation of the method and, for illustrative purposes, calculate explicitly the Backus-Gilbert solutions in a standard Hausdorff problem for some standard choices of the parameters therein.

Let $v_1, \ldots, v_n$ be functions defined on Ω. Let A and B be the linear operators defined by

$$Au = \left(\int_\Omega u(y)g_1(y)dy, \ldots, \int_\Omega u(y)g_n(y)dy \right),$$

(u is in an appropriate function space) and

$$B_v(\mu) = u^n = \sum_{j=1}^n \mu_j v_j \quad (\mu = (\mu_1, \ldots, \mu_n) \in \mathbf{R}^n).$$

The Backus-Gilbert method is to find the functions $v = (v_1, \ldots, v_n)$ making the composition $B_v A$ as close to the identity mapping as possible. The approximation $B_v A u$ of u is related to the original function u by the equation

$$B_v Au(x) = \int_\Omega \left(\sum_{j=1}^n g_j(y)v_j(x) \right) u(y)dy. \tag{3.1}$$

The resolvent function (also called the averaging kernel) is thus

$$\hat{\delta}_v(x,y) = \sum_{j=1}^n g_j(y)v_j(x).$$

If $\hat{\delta}_v(x,y)$ approximates the Delta function $\delta(x-y)$ (in an appropriate sense) then we hope that $B_v A$ is close to the identity mapping. The Backus-Gilbert method aims to minimize the "spread" of $\hat{\delta}_v(x,y)$ (that is, to maximize the resolving power). For this purpose, Backus and Gilbert (see [KSB], [HSo], [BG]) proposed a measure of this spread of $\hat{\delta}_v(x,y)$ (at each value of x) by the convolution-type integral

$$S_x^n(v) = \int_\Omega |x-y|^2 [\hat{\delta}_v(x,y)]^2 dy.$$

In fact, we shall find v such

$$B_v Au = u \ \text{ for } u \equiv const \tag{3.2}$$

and that

$$|B_v Au(x) - u(x)| \text{ is as small as possible for each } x \in \Omega.$$

One has in view of (3.1), (3.2)

$$\sum_{j=1}^n v_j(x) \int_\Omega g_j(y)dy = 1 \tag{3.3}$$

and

$$B_v Au(x) - u(x) = B_v Au(x) - u(x)B_v A1$$
$$= \int_\Omega (u(y) - u(x))\hat{\delta}_v(x,y)dy.$$

Using standard arguments of functional analysis, we get, from the latter equality, the "spread" of $\hat{\delta}_v(x,y)$ as

$$\sup_{u \in L_x} |B_v Au(x) - u(x)|^2 = \int_\Omega |x-y|^2 [\hat{\delta}_v(x,y)]^2 dy \equiv S_x^n(v)$$

where

$$L_x = \left\{ u \in C(\overline{\Omega}) : \int_\Omega \frac{|u(y) - u(x)|^2}{|y-x|^2} dy \le 1 \right\}.$$

Hence, one has

$$\min_v S_x^n(v) = \min_v \sup_{u \in L_x} |B_v Au(x) - u(x)|^2$$

where v satisfies (3.3). For fixed n, the functions v_i^n, $i = 1, ..., n$, $(v^n = (v_1^n, ..., v_n^n))$ satisfying the foregoing minimization procedure are called the Backus-Gilbert functions (see Sect. 3.1. for details). The solution u of the problem (MP) will be approximated by

$$u^n = B_{v^n} Au = \sum_{j=1}^{n} \mu_j v_j^n$$

(called Backus-Gilbert solution).

Now, we present an example in which the Backus-Gilbert functions are calculated explicity. We consider the Hausdorff moment problem on $\Omega = (0, 1)$, that is $g_j(y) = y^{j-1}$ for $y \in (0, 1)$, $j \in \mathbf{N}$. Accordingly, the functional S_x^n and the constraint (3.3) become

$$S_x^n(v) = \int_0^1 (x - y)^2 \left(\sum_{j=1}^{n} v_j y^{j-1} \right)^2 dy$$

and

$$\sum_{j=1}^{n} \frac{1}{j} v_j = 1.$$

Let

$$\gamma = (1, 1/2, ..., 1/n)^T \in \mathbf{R}^n$$

and let $G(x)$ be the $n \times n$ matrix with entries

$$G_{ij}(x) = \int_0^1 (x - y)^2 y^{i-1} y^{j-1} dy$$

$$= \frac{1}{i+j-1} x^2 - \frac{2}{i+j} x + \frac{1}{i+j+1} \quad (i, j \in \{1, ..., n\}).$$

Using the Lagrange multiplier rule, we see that the constrained minimization problem

$$\min_v S_x^n(v)$$

for v satisfying (3.3) is equivalent to the problem of finding $(v, \lambda) \in \mathbf{R}^{n+1}$ such that

$$\begin{cases} G(x)v + \lambda v = 0 \\ v^T \gamma = 1 \end{cases}$$

that is, (v, λ) is the (unique) solution of the linear system

$$\begin{pmatrix} G(x) & \gamma \\ \gamma^T & 0 \end{pmatrix} \begin{pmatrix} v \\ \lambda \end{pmatrix} = \begin{pmatrix} 0 \\ 1 \end{pmatrix}$$

(cf. [KSB], [HSo]). Also, according to Lemma 2.2 in [KSB], for each $n \in \mathbf{N}$, the Backus-Gilbert functions $v_j(x) = v_j^n(x)$ $(j = 1, ..., n)$ are rational functions whose numerators and denominators are polynomials of degrees at most $2(n - 1)$. The denominators have no zeros in $(0, 1)$. Therefore, the approximate solutions $u^n(x)$ are always rational functions with no singularities on $(0, 1)$. Using the latter linear system, together with interpolation and other numerical methods, one can find the functions v_j^n. For example, in the case $n = 4$, the Backus-Gilbert solutions are:

$$
\begin{cases}
v_1^4(x) = \frac{1}{2} \frac{882x^6 - 3528x^5 + 5852x^4 - 5208x^3 + 2625x^2 - 700x + 75}{441x^6 - 1323x^5 + 1603x^4 - 1001x^3 + 343x^2 - 63x + 6} \\[2mm]
v_2^4(x) = 7 \frac{126x^5 - 504x^4 + 836x^3 - 654x^2 + 225x - 25}{441x^6 - 1323x^5 + 1603x^4 - 1001x^3 + 343x^2 - 63x + 6} \\[2mm]
v_3^4(x) = \frac{21}{2} \frac{126x^4 - 504x^3 + 566x^2 - 224x + 25}{441x^6 - 1323x^5 + 1603x^4 - 1001x^3 + 343x^2 - 63x + 6} \\[2mm]
v_4^4(x) = 126 \frac{14x^3 - 21x^2 + 9x - 1}{441x^6 - 1323x^5 + 1603x^4 - 1001x^3 + 343x^2 - 63x + 6}.
\end{cases}
$$

For further illustrations, let us calculate the Backus-Gilbert approximation $u^4(x)$ for the (original) function

$$u(x) = e^x, \ x \in (0, 1).$$

The moments are, in this case,

$$\mu_1 = e - 1, \mu_2 = 1, \mu_3 = e - 2,$$

$$\mu_4 = -2e + 6, \mu_5 = 9e - 24, \mu_6 = -44e + 120, \ldots$$

and the Backus-Gilbert approximation is the rational function

$$
\begin{aligned}
u^4(x) = \frac{1}{2} & [882(e - 1)x^6 + (5292 - 3528e)x^5 + (8498e - 18200)x^4 \\
& + (59248 - 22848e)x^3 + (5095e - 673052)x^2 \\
& + (26866 - 9940e)x - 2987 + 1104e] \\
& \times (441x^4 - 1323x^5 + 1603x^4 - 1001x^3 + 343x^2 - 63x + 6)^{-1}.
\end{aligned}
$$

The remainder of the chapter consists of two sections. In Section 3.2, we extend the concept of Backus-Gilbert solution (henceforth, also written BG solution, for short) to the case where $q \in (1, \infty)$ and the exponent of the weight function is a positive real number (Sect. 3.2.1). In Sect. 3.2.2, the stability of BG solutions is proved. Section 3.3 is devoted to our regularization of (MP) via the Backus-Gilbert solutions. Some preliminary lemmas are presented in Sect. 3.3.1. For the proofs of these lemmas, we have used ideas from [KSB]. Sect. 3.3.2 contains the main results of the chapter, namely Theorem 2 and Theorem 3, which give stability estimates for the BG solutions. The paper concludes with Theorems 3.6 and 3.7, which extend Theorem 3.1 and Theorem 4.2 of [KSB]. Compare also [AGV1].

3.2 Backus-Gilbert solutions and their stability

3.2.1 Definition of the Backus-Gilbert solutions

We consider problem (MP) with the following assumptions :

The vector space spanned by $g_1, g_2, \ldots$ is dense in $L^q(\Omega)$. \hfill (3.4)

$$g_1, ..., g_n \text{ are linearly independent for all } n \in N. \tag{3.5}$$

The Backus-Gilbert method consists in approximating the solution u of (MP) by an appropriate sequence of functions $u^n = u^n(., \mu)$, stable with respect to variations in μ, that are linear combinations of some predetermined functions called Backus-Gilbert basis functions (see [KSB]).

For $x \in \mathbf{R}^d$, the values of the Backus-Gilbert basis functions at x are defined by the following minimization procedure :

$$min_{v \in L_n} S_x^n(v) \tag{3.6}$$

where

$$L_n = \{v \in \mathbf{R}^n \; : \; \sum_{j=1}^{n} v_j \int_{\Omega} g_j(y) dy = 1\}$$

and

$$S_x^n(v) = S_x(v) = \int_{\Omega} |x - y|^\beta \left| \sum_{j=1}^{n} v_j g_j(y) \right|^q dy$$

Here $n \in N$, and β is a fixed positive number (when $q = 2$ and $\beta = 2k$, $k \in N$, this reduces to the problem considered in [KSB]).

The following proposition guarantees the existence and uniqueness of the Backus-Gilbert (BG) basis functions.

Proposition 3.1. *For each $x \in \mathbf{R}^d$, the functional S_x is continuous, strictly convex and coercive on $\mathbf{R}^n$ in the sense that there exists $C > 0$ such that*

$$S_x(v) \geq C|v|^q, \quad \forall v \in \mathbf{R}^n.$$

Consequently, the minimization problem (3.6) has a unique solution.

Proof.

Let $v_m \to v$ in $\mathbf{R}^n$. Then

$$\sum_{j=1}^{n} v_{mj} g_j \to \sum_{j=1}^{n} v_j g_j \text{ in } L^q(\Omega)$$

and thus

$$\left| \sum_{j=1}^{n} v_{mj} g_j \right|^q \to \left| \sum_{j=1}^{n} v_j g_j \right|^q \text{ in } L^1(\Omega).$$

Therefore,

$$\int_{\Omega} |x - y|^\beta \left| \sum_{j=1}^{n} v_{mj} g_j(y) \right|^q dy \to \int_{\Omega} |x - y|^\beta \left| \sum_{j=1}^{n} v_j g_j(y) \right|^q dy \; (m \to \infty).$$

This means

$$S_x(v_m) \to S_x(v) \quad (m \to \infty).$$

Hence S_x is continuous on $\mathbf{R}^n$.

For the proof of the strict convexity of S_x, observe first that the function $t \mapsto t^q$ from $[0, \infty)$ to $[0, \infty)$ is strictly convex.

For $v, w \in \mathbf{R}^n, y \in \Omega$, we put

$$(v, w) = \sum_{j=1}^{n} v_j w_j \text{ and } g(y) = (g_1(y), ..., g_n(y)).$$

Now, let $v, w \in \mathbf{R}^n$, $\lambda, \mu \geq 0$ and $\lambda + \mu = 1$. We have

$$\begin{aligned}
S_x(\lambda v + \mu w) &= \int_\Omega |x - y|^\beta |(\lambda v + \mu w, g(y))|^q dy \\
&\leq \lambda \int_\Omega |x - y|^\beta |(v, g(y))|^q dy \\
&\quad + \mu \int_\Omega |x - y|^\beta |(w, g(y))|^q dy \\
&= \lambda S_x(v) + \mu S_x(w).
\end{aligned} \tag{3.7}$$

Suppose $0 < \lambda, \mu < 1$ and that equality holds in (3.7). Then we have

$$\begin{aligned}
|(\lambda v + \mu w, g(y))|^q &= \{\lambda |(v, g(y))| + \mu |(w, g(y))|\}^q \\
&= \lambda |(v, g(y))|^q + \mu |(w, g(y))|^q \quad a.e., \ y \in \Omega.
\end{aligned}$$

It follows from the first equality that $(v, g(y))$ and $(w, g(y))$ have the same sign. From the second equality the strict convexity of the function t^q implies

$$|(v, g(y))| = |(w, g(y))|.$$

Then

$$(v, g(y)) = (w, g(y)) \quad a.e., \ y \in \Omega.$$

This and the linear independence of $g_1, ..., g_n$ imply

$$v = w \tag{3.8}$$

and the strict convexity of S_x follows.

We can also conclude from the linear independence of $g_1, ..., g_n$ that

$$S_x(v) > 0 \text{ if } v \in \mathbf{R}^n \setminus \{(0, ..., 0)\}.$$

Hence $C \equiv \min_{|v|=1} S_x(v) > 0$. On the other hand, it is easy to prove that

$$S_x(\lambda v) = |\lambda|^q S_x(v) \quad \forall v \in \mathbf{R}^n, \ \forall \lambda \in \mathbf{R}. \tag{3.9}$$

Then, for $v \neq (0, ..., 0)$,

$$|v|^{-q} S_x(v) = S_x(|v|^{-1}v) \geq C,$$

i.e.

$$S_x(v) \geq C|v|^q, \qquad \forall v \in \mathbf{R}^n.$$

This implies $S_x(v) \to \infty$ as $|v| \to \infty$.

Now, since L_n is closed in $\mathbf{R}^n$ and by the continuity of S_x, the foregoing relation implies the existence of a solution of (3.6). By the strict convexity of S_x, we get the uniqueness of the solution of (3.6). This completes the proof of Proposition 3.1.

From Proposition 3.1, we can define a function from $\mathbf{R}^d$ to $\mathbf{R}^n$:

$$x \mapsto v(x) = v^n(x)$$

such that $v^n(x) \in L_n$ and

$$S_x^n(v^n(x)) = \min_{w \in L_n} S_x^n(w). \tag{3.10}$$

Definition 3.1. *(a) Let $v^n(x) = (v_1^n(x), ..., v_n^n(x)) = (v_1(x), ..., v_n(x))$, $x \in \mathbf{R}^d$ be defined by (3.10). We call $v_1^n, ..., v_n^n$ the Backus-Gilbert functions of order n.*
(b) For $\mu_1, ..., \mu_n \in \mathbf{R}$,

$$u^n = u^n(\mu_1, ..., \mu_n) = \sum_{j=1}^n \mu_j v_j^n$$

is called the Backus-Gilbert solution corresponding to $\mu_1, ..., \mu_n$ (and $g_1, ..., g_n$).

The following proposition gives us an elementary property of the BG basis functions.

Proposition 3.2. *The Backus-Gilbert basis functions are continuous on $\mathbf{R}^d$.*

Proof. Let $v_1, ..., v_n$ be the BG basis functions and let $x_m \to x$ in $\mathbf{R}^d$. Remark first that

$$S_{x_m} \to S_x \text{ uniformly on } A = \{w \in \mathbf{R}^n : |w| = 1\}. \tag{3.11}$$

where $|.|$ is Euclidean norm in $\mathbf{R}^n$. Indeed, for $w \in A$, we have

$$|S_{x_m}(w) - S_x(w)| \leq \int_\Omega ||x_m - y|^\beta - |x - y|^\beta| \left(\sum_{j=1}^n |w_j||g_j(y)| \right)^q dy$$

$$\leq \left(\sup_{y \in \Omega} ||x_m - y|^\beta - |x - y|^\beta| \right) \int_\Omega \left(\sum_{j=1}^n |g_j(y)| \right)^q dy.$$

Using the inequalities

$$|x^\beta - y^\beta| \le |x - y|^\beta \quad (x, y \ge 0, \ 0 < \beta < 1)$$

and

$$|x^\beta - y^\beta| \le \beta |x - y| \{\max(x, y)\}^{\beta-1} \quad (x, y \ge 0, \ \beta \ge 1),$$

we can prove directly that

$$\sup_{y \in \Omega} ||x_m - y|^\beta - |x - y|^\beta| \to 0 \quad (m \to \infty) \tag{3.12}$$

and then obtain (3.11).

It can be shown as in the proof of Proposition 3.1 that there exists $C > 0$ so that

$$S_x(w) > C, \qquad \forall w \in A.$$

From (3.11), there is an $m_0 \in \mathbf{N}$ such that

$$S_{x_m}(w) > C, \quad \forall w \in A, \quad \forall m \ge m_0.$$

By (3.9),

$$S_{x_m}(w) \ge C|w|^q \qquad \forall w \in \mathbf{R}^n, \quad \forall m \ge m_0.$$

Now, fix $w_0 \in L_n$. We have, for $m \ge m_0$,

$$C|v(x_m)|^q \le S_{x_m}(v(x_m)) \le S_{x_m}(w_0).$$

Since $\sup_{y \in \Omega, m \ge m_0} |x_m - y|^\beta < \infty$, this means that $\{v(x_m)\}_m$ is a bounded sequence in $\mathbf{R}^n$. Suppose $\bar{v}$ is a limit point of $(v(x_n))$, i.e.,

$$v(x_{m_k}) \to \bar{v} \quad (k \to \infty)$$

for a subsequence (x_{m_k}) of (x_m). We have, for $w \in L_n$, $k \ge 1$,

$$
\begin{aligned}
S_{x_{m_k}}(v(x_{m_k})) &= \int_\Omega |x_{m_k} - y|^\beta |(v(x_{m_k}), g(y))|^q dy \\
&\le \int_\Omega |x_{m_k} - y|^\beta |(w, g(y))|^q dy. \\
&= S_{x_{m_k}}(w). \tag{3.13}
\end{aligned}
$$

From (3.12) and the fact that

$$|(v(x_{m_k}), g)|^q \to |(\bar{v}, g)|^q \quad \text{in } L^1(\Omega)$$

we obtain

$$\int_\Omega |x_{m_k} - y|^\beta |(v(x_{m_k}), g(y))|^q dy \to \int_\Omega |x - y|^\beta |(\bar{v}, g(y))|^q dy \quad (k \to \infty).$$

Similarly,

$$\int_\Omega |x_{m_k} - y|^\beta |(w, g(y))|^q dy \to \int_\Omega |x - y|^\beta |(\bar{v}, g(y))|^q dy.$$

Letting $k \to \infty$ in (3.13), we obtain

$$S_x(\overline{v}) \leq S_x(w).$$

Since this holds for every $w \in L_n$, we have, by the uniqueness of the solution of (3.6), $\overline{v} = v(x)$.

Therefore, $v(x)$ is the unique limit point of the bounded sequence $\{v(x_m)\}$. Hence $v(x_m) \to v(x)$ $(m \to \infty)$ and Proposition 3.2 is proved.

3.2.2 Stability of the Backus-Gilbert solutions

In view of Definition 3.1 (b), we can prove for a fixed n, that the BG solutions depend continuously on $\mu_1, ..., \mu_n$. We give here a somewhat stronger stability property of the BG solutions.

Theorem 3.3. (a) Let $(g_1^\lambda, ..., g_n^\lambda)_{\lambda \in \mathbf{N}}$ be a sequence in $[L^q(\Omega)]^n$ such that

$$g_j^\lambda \to g_j \quad in \ \ L^q(\Omega), j = 1, ..., n.$$

Then $g_1^\lambda, ..., g_n^\lambda$ are linearly independent for all λ sufficiently large, and

$$v_j^\lambda \to v_j \quad uniformly \ on \ compact \ subsets \ of \ \mathbf{R}^d,$$

where $v_1^\lambda, .., v_n^\lambda$ (resp. $v_1, ..., v_n$) denote the BG solutions corresponding to $g_1^\lambda, ..., g_n^\lambda$ (resp. $g_1, ..., g_n$).

(b) Suppose furthermore that $(\mu_1^\lambda, ..., \mu_n^\lambda)_{\lambda \in \mathbf{N}}$ is a sequence in $\mathbf{R}^n$ and that

$$(\mu_1^\lambda, ..., \mu_n^\lambda) \to (\mu_1, ..., \mu_n) \quad in \ \mathbf{R}^n \ \ (\lambda \to \infty).$$

Let u_λ^n (resp. u^n) be the BG solution corresponding to $\mu_1^\lambda, ..., \mu_n^\lambda$ and $g_1^\lambda, ..., g_n^\lambda$ (resp. $\mu_1, ..., \mu_n$ and $g_1, ..., g_n$). Then $u_\lambda^n \to u^n$ uniformly on compact subsets of $\mathbf{R}^d$. In particular, $u_\lambda^n \to u^n$ uniformly on $\overline{\Omega}$.

Proof. (a) Let V be the subspace of $L^q(\Omega)$ generated by $g_1, ..., g_n$. Since $\dim V < \infty$, there exists an inner product $(.,.)_V$ in V, and V admits a topological complement in $L^q(\Omega)$, that is, there exists a bounded linear mapping

$$P : L^q(\Omega) \to V$$

such that

$$P(g) = g, \qquad \forall g \in V.$$

We claim that $P(g_1^\lambda), ..., P(g_n^\lambda)$ are linearly independent for λ sufficiently large. This will hold if the Gram determinant of $P(g_1^\lambda), ..., P(g_n^\lambda)$ is different from 0. Since

$$P(g_j^\lambda) \to P(g_j) = g_j \quad in \ \ V \ \ (\lambda \to \infty), \ j = 1, 2, .., n,$$

we see that

$$\det[(P(g_i^\lambda), P(g_j^\lambda))_V]_{i,j=1,...,n} \to \det[(g_i, g_j)_V]_{i,j=1,...,n} \quad as \ \lambda \to \infty.$$

But $\det[(g_i, g_j)_V]_{i,j=1,...,n} \neq 0$. Then, for λ sufficiently large,

$$\det[(P(g_i^\lambda), P(g_j^\lambda))_V]_{i,j=1,...,n} \neq 0$$

This means the linear independence of $P(g_1^\lambda), ..., P(g_n^\lambda)$, and then of $g_1^\lambda, ..., g_n^\lambda$ for λ sufficiently large, e.g., for $\lambda \geq \lambda_0$ ($\lambda_0 \in \mathbf{N}$). Now, we prove that the set

$$\{ v^\lambda = (v_1^\lambda, ..., v_n^\lambda) \; : \; \lambda \geq \lambda_0 \}$$

is equicontinuous on each compact subset K of $\mathbf{R}^d$.

Suppose by contradiction that there exist a compact set K in $\mathbf{R}^d$, $\epsilon_0 > 0$ and sequences $(\lambda_m), (x_m), (y_m)$ such that $x_m, y_m \in K$,

$$|x_m - y_m| < 1/m \text{ and } |v^{\lambda_m}(x_m) - v^{\lambda_m}(y_m)| > \epsilon_0, \; \forall m \in \mathbf{N}. \tag{3.14}$$

If $\lambda_m = \lambda$ for all m sufficiently large then $v^{\lambda_m} = v^\lambda$ and by the uniform continuity of v^λ on K (Proposition 3.2), the conditions in (3.14) cannot hold at the same time. Therefore, by extracting a subsequence of $\{\lambda_m\}$, we can assume that $\lambda_m \to \infty$ as $m \to \infty$. Similarly, we can also assume that $x_m \to x \in K$ ($m \to \infty$). By (3.14), the latter assumption implies $y_m \to x$.

By the definition of v^λ, we have

$$S_x^\lambda(v^\lambda(x)) = \min_{w \in L_n^\lambda} S_x^\lambda(w),$$

where

$$L_n^\lambda = \left\{ w = (w_1, .., w_n) \in \mathbf{R}^n \; : \; \sum_{j=1}^n w_j \left(\int_\Omega g_j^\lambda(y) dy \right) = 1 \right\}$$

and

$$S_x^\lambda(w) = \int_\Omega |x - y|^\beta \left| \sum_{j=1}^n w_j g_j^\lambda(y) \right|^q dy.$$

We show that

$$S_{x_m}^{\lambda_m}(w) \to S_x(w) \text{ uniformly on } A \text{ (A is defined in (3.11)).} \tag{3.15}$$

We have, for $w \in A$,

$$|S_{x_m}^{\lambda_m}(w) - S_x(w)| \leq I_1^m + I_2^m$$

with

$$I_1^m = \int_\Omega |x_m - y|^\beta \left| |\sum_{j=1}^n w_j g_j^{\lambda_m}(y)|^q - |\sum_{j=1}^n w_j g_j(y)|^q \right| dy,$$

$$I_2^m = \int_\Omega \left| |x_m - y|^\beta - |x - y|^\beta \right| \left| \sum_{j=1}^n w_j g_j(y) \right|^q dy.$$

By direct computations, it follows that

$$I_1^m \leq q \int_\Omega |x_m - y|^\beta \left| \sum_{j=1}^n w_j \left[g_j^{\lambda_m}(y) - g_j(y) \right] \right| \times$$

$$\times \left[\max \left\{ \left| \sum_{j=1}^n w_j g_j^{\lambda_m}(y) \right|, \left| \sum_{j=1}^n w_j g_j(y) \right| \right\} \right]^{q-1} dy$$

$$\leq q \int_\Omega |x_m - y|^\beta \left(\sum_{j=1}^n |g_j^{\lambda_m}(y) - g_j(y)| \right) \left(\sum_{j=1}^n |g_j^{\lambda_m}| + \sum_{j=1}^n |g_j(y)| \right)^{q-1} dy$$

$$\leq q \sup_{y \in \Omega, m \geq 1} |x_m - y|^\beta \left[\int_\Omega \left(\sum_{j=1}^n |g_j^{\lambda_m}(y) - g_j(y)| \right)^q dy \right]^{\frac{1}{q}} \times$$

$$\times \left[\int_\Omega \left(\sum_{j=1}^n |g_j^{\lambda_m}(y)| + \sum_{j=1}^n |g_j(y)| \right)^q dy \right]^{\frac{q-1}{q}}$$

But $\sup_{y \in \Omega, m \geq 1} |x_m - y|^\beta < \infty$, the sequence

$$\left\{ \int_\Omega \left(\sum_{j=1}^n |g_j^{\lambda_m}(y)| + \sum_{j=1}^n |g_j(y)| \right)^q dy \right\}_{m \in \mathbf{N}}$$

is bounded and

$$\int_\Omega \left(\sum_{j=1}^n |g_j^{\lambda_m}(y) - g_j(y)| \right)^q dy \to 0 \quad \text{as} \quad m \to \infty.$$

Hence $I_1^m \to 0$ uniformly on A ($m \to \infty$). On the other hand, as in (3.11), we can show that $I_2^m \to 0$ uniformly on A. These imply (3.15).

From (3.15) and the fact that $S_x(w) > C$ ($\forall w \in A$) for some $C > 0$, we have an $m_0 \in N$ such that

$$S_{x_m}^{\lambda_m}(w) > C, \ \forall w \in A, \ \ \forall m \geq m_0.$$

Therefore,

$$S_{x_m}^{\lambda_m}(w) > C|w|^q, \qquad \forall w \in \mathbf{R}^n, \ \ \forall m \geq m_0.$$

Now, fix an element $w \in L_n$ and define

$$\gamma = \left(\int_\Omega g_1, ..., \int_\Omega g_n \right), \quad \gamma_m = \left(\int_\Omega g_1^{\lambda_m}, ..., \int_\Omega g_n^{\lambda_m} \right)$$

$$w_m = (w, \gamma_m)^{-1} w \quad (m \geq 1). \tag{3.16}$$

Since $\gamma_m \to \gamma$ in $\mathbf{R}^n$, w_m is well defined and $w_m \in L_n^{\lambda_m}$ for all large m. Then

$$
\begin{aligned}
C|v^{\lambda_m}(x_m)|^q &\le S_{x_m}^{\lambda_m}(v^{\lambda_m}(w_m)) \\
&\le S_{x_m}^{\lambda_m}(w_m) \\
&\le \int_\Omega |x_m - y|^\beta \left(\sum_{j=1}^n |w_{mj}||g_j^{\lambda_m}(y)| \right)^q dy \\
&\le \sup_{y \in \Omega, m \ge 1} |x_m - y|^\beta \max_{1 \le j \le n} |w_{mj}|^q \int_\Omega \left(\sum_{j=1}^n |g_j^{\lambda_m}(y)| \right)^q dy.
\end{aligned}
$$

As shown above, the set

$$
\left\{ \int_\Omega \left(\sum_{j=1}^n |g_j^{\lambda_m}(y)| \right)^q dy \right\}_{m \ge 1} \qquad \text{is bounded.}
$$

Since $(w, \gamma_m) \to (w, \gamma) = 1 \ (m \to \infty)$, we have

$$
w_m \to w \quad \text{in} \ \mathbf{R}^n.
$$

Hence, $\{\max\{ |w_{mj}| : 1 \le j \le n \}\}_{m \ge 1}$ is also bounded.

It follows that the sequence $\{v^{\lambda_m}(x_m)\}_m$ is bounded in $\mathbf{R}^n$. Similarly, $\{v^{\lambda_m}(y_m)\}_m$ is bounded. Hence, by passing to a subsequence, we can assume that

$$
v^{\lambda_m}(x_m) \to v_0, \quad v^{\lambda_m}(y_m) \to v_1 \quad (m \to \infty). \tag{3.17}
$$

Now, let $w \in L_n$ and define w_m from w by (3.16). We have, for all m sufficiently large

$$
S_{x_m}^{\lambda_m}(v^{\lambda_m}(x_m)) \le S_{x_m}^{\lambda_m}(w_m). \tag{3.18}
$$

However,

$$
|S_{x_m}^{\lambda_m}(v^{\lambda_m}(x_m)) - S_x(v_0)| \le |S_{x_m}^{\lambda_m}(v^{\lambda_m}(x_m)) - S_{x_m}^{\lambda_m}(v_0)| + |S_{x_m}^{\lambda_m}(v_0) - S_x(v_0)|.
$$

Hence

$$
\begin{aligned}
|S_{x_m}^{\lambda_m}(v^{\lambda_m}(x_m)) &- S_{x_m}^{\lambda_m}(v_0)| \le \\
&\le \int_\Omega |x_m - y|^\beta \left| |\sum_{j=1}^n [v^{\lambda_m}(x_m)]_j g_j(y)|^q - |\sum_{j=1}^n v_{0j} g_j(y)|^q \right| dy.
\end{aligned}
$$

Since

$$
\sum_{j=1}^n [v^{\lambda_m}(x_m)]_j g_j \to \sum_{j=1}^n v_{0j} g_j \quad \text{in} \ L^q(\Omega),
$$

we have

$$
|S_{x_m}^{\lambda_m}(v^{\lambda_m}(x_m)) - S_{x_m}^{\lambda_m}(v_0)| \to 0 \quad (m \to \infty).
$$

As in the proof of (3.15), we can show that

$$S_{x_m}^{\lambda_m}(v_0) \to S_x(v_0) \quad (m \to \infty).$$

Then

$$S_{x_m}^{\lambda_m}(v^{\lambda_m}(x_m)) \to S_x(v_0) \quad \text{as } m \to \infty.$$

Similarly, from $w_m \to w$ in $\mathbf{R}^n$, we obtain

$$S_{x_m}^{\lambda_m}(w_m) \to S_x(w) \quad \text{as } m \to \infty.$$

Letting $m \to \infty$ in (3.18), we have $S_x(v_0) \leq S_x(w)$. Since this holds for every $w \in L_n$, we have, via Proposition 3.1,

$$v_0 = v(x). \tag{3.19}$$

Similarly, $v_1 = v(x)$; i.e., $v_0 = v_1$. This and (3.17) contradict (3.14). Hence $\{v^\lambda\}_{\lambda \geq \lambda_0}$ is equicontinuous on every compact subset of $\mathbf{R}^d$.

Now, let $x \in \mathbf{R}^d$. As in the above proof, we can show that there exist $C > 0, \lambda_0 \in \mathbf{N}$ such that

$$S_x^\lambda(v^\lambda(x)) \geq C \left| v^\lambda(x) \right|^q, \quad \forall \lambda \geq \lambda_0.$$

Repeating the proof of the boundedness of $\{v^{\lambda_m}(x_m)\}_m$ we have that $\{v^\lambda(x)\}_\lambda$ is bounded in $\mathbf{R}^n$. Suppose $\bar{v}$ is a limit point of this sequence. By a proof similar to that of (3.19), we obtain $\bar{v} = v(x)$. Therefore $v(x)$ is the unique limit point of the bounded sequence $\{v^\lambda(x)\}_\lambda$. It follows that

$$\lim_{\lambda \to \infty} v^\lambda(x) = v(x), \quad x \in \mathbf{R}^d. \tag{3.20}$$

Let K be a compact subset of $\mathbf{R}$. Ascoli's theorem shows that the set

$$\left\{ v^\lambda|_K \ : \ \lambda \geq \lambda_0 \right\}$$

is relatively compact in $C(K)$. (3.20) also implies that $v|_K$ is the unique limit point of $\{v^\lambda|_K\}_\lambda$ in $C(K)$. Hence $v^\lambda|_K \to v|_K$ in $C(K)$, i.e., $v^\lambda \to v$ uniformly on compact subsets of $\mathbf{R}^d$.

(b) From (a) and the definitions of u_λ^n, u^n, we have

$$u_\lambda^n = \sum_{j=1}^n \mu_j^\lambda v_j^\lambda \to \sum_{j=1}^n \mu_j v_j = u^n$$

uniformly on compact subsets of $\mathbf{R}^d$ as $\lambda \to \infty$. This completes the proof of Theorem 3.3.

3.3 Regularization via Backus-Gilbert solutions

In view of the discussion in Introduction of the present chapter, (MP) is often ill-posed, and a regularization is in order. We shall present here a regularization method based on the Backus-Gilbert solutions. We show that with an appropriate order n, the BG solution $u^n(\mu)$ can be regarded as a stabilized approximate solution of (MP). This is the content of Section 3.2.2. In Section 3.2.1, we prove some preparatory lemmas needed for the analysis in Section 3.2.2.

3.3.1 Definitions and notations

For $n \in \mathbf{N}$, let V_n be the subspace of $L^q(\Omega)$ generated by $g_1, ..., g_n$. Put

$$W = \left\{ v \in L^q(\Omega) \ : \ \int_\Omega v = 1 \right\} \quad \text{and} \quad W_n = W \cap V_n \ (n \geq 1).$$

For $x \in \mathbf{R}^d$, $n \geq 1$, put

$$\epsilon_n(x) = \min_{v \in W_n} \int_\Omega |x - y|^\beta |v(y)|^q dy = \min_{w \in L_n} S_x(w).$$

By Definition 3.1 (a),

$$\epsilon_n(x) = S_x(v^n(x)). \tag{3.21}$$

We also need some geometrical properties of Ω.

As in [KSB], we say that Ω satisfies a cone condition if

$$\forall x \in \partial\Omega, \exists \lambda_x \in (0,1), \exists a_x \in \mathbf{R}^d \text{ such that } |a_x| = 1 \text{ and}$$

$$C(x, a_x, \lambda_x) \subset \Omega.$$

Here

$$C(x, a, \lambda) = \left\{ y \in \mathbf{R}^d : \ |x - y| < \lambda \text{ and } a^T(x - y) > (1 - \lambda)|x - y| \right\}.$$

We say that Ω satisfies a uniform cone condition if there exists a $\lambda \in (0,1)$ satisfying

$$\forall x \in \overline{\Omega}, \exists a_x \in \mathbf{R}^d \text{ such that } |a_x| = 1 \text{ and } C(x, a_x, \lambda) \subset \Omega.$$

The following proposition presents some properties of ϵ_n.

Proposition 3.3. *(a)* $\{\epsilon_n\}_{n \in \mathbf{N}}$ *is a nonincreasing sequence of continuous functions on* $\mathbf{R}^d$.

(b) If $\beta \geq (q-1)d$ *then* $\epsilon_n \to 0$ *uniformly on compact subsets of* Ω. *If* Ω *satisfies a cone condition, then* $\epsilon_n \to 0$ *uniformly on* $\overline{\Omega}$.

(c) If $\beta < (q-1)d$ *then* ϵ_n *converges to the function*

$$\epsilon : x \longmapsto \left(\int_\Omega |x - y|^{\beta/(1-q)} dy \right)^{1-q}$$

uniformly on compact subsets of $\mathbf{R}^d$. *Consequently, there exists an* $m_0 > 0$ *such that*

$$\epsilon_n(x) \geq m_0, \qquad \forall x \in \overline{\Omega}, \qquad \forall n \geq 1.$$

Proof.

(a) It is clear that the sequence $(\epsilon_n(x))_n$ $(x \in \mathbf{R}^d)$ is nonincreasing. The continuity of ϵ_n is a consequence of (3.21) and of Proposition 2.

(b) For our proof, we use some ideas in the proof of Lemma 3.2 of [KSB]. Let us recall first a well known result concerning integrals : For $\lambda \in (0,1), a \in \mathbf{R}^d, |a| = 1$,

$$\begin{cases} \displaystyle\int_{C(0,a,\lambda)} |x|^{-\gamma} dx = \infty \Leftrightarrow \gamma \geq d \\ \displaystyle\int_{B(0,1)} |x|^{-\gamma} dx = \infty \Leftrightarrow \gamma \geq d, \end{cases} \tag{3.22}$$

where $B(0,1)$ is the unit ball in $\mathbf{R}^d$.

Suppose now that $\zeta = \beta/(q-1) \geq d$. Fix $x \in \Omega$ and choose, for each $n \in \mathbf{N}$, a function $\bar{v}_n$ satisfying :

$$\begin{cases} \bar{v}_n \in C(\overline{\Omega}) \\ 0 \leq \bar{v}_n(y) \leq |x-y|^{-\zeta}, \quad \forall y \in \overline{\Omega} \\ \bar{v}_n(y) = |x-y|^{-\zeta} \quad \text{if } |x-y| \geq 1/n. \end{cases}$$

Since

$$v_n = \left(\int_\Omega \bar{v}_n\right)^{-1} \bar{v}_n \in W_n,$$

we have

$$0 \leq \epsilon_n(x) \leq \int_\Omega |x-y|^\beta |v_n(y)|^q dy$$

$$\leq \left(\int_\Omega \bar{v}_n\right)^{-q} \int_\Omega |x-y|^{\beta-\zeta(q-1)} \bar{v}_n(y) dy$$

$$= \left(\int_\Omega \bar{v}_n\right)^{1-q}.$$

Since $x \in \Omega$, $B(x,r) \subset \Omega$ for some $r > 0$, and by (3.22),

$$\int_\Omega |x-y|^{-\zeta} dy = \int_{B(x,r)} |x-y|^{-\zeta} dy = \infty.$$

and then

$$\liminf_{n\to\infty} \int_\Omega \bar{v}_n \geq \liminf_{n\to\infty} \int_{\{ y\in\Omega \,:\, |x-y|\geq 1/n \}} |x-y|^{-\zeta} dy$$

$$= \int_\Omega |x-y|^{-\zeta} dy = \infty.$$

It follows that

$$\lim_{n\to\infty} \epsilon_n(x) = 0 \quad (x \in \Omega). \tag{3.23}$$

By (a) and Dini's theorem, we see that $\epsilon_m \to 0$ uniformly on compact subsets of Ω.

Now suppose furthermore that Ω satisfies a cone condition. Then (3.23) holds also for $x \in \partial\Omega$. Indeed, let $x \in \partial\Omega$. Since $C(x,a,\lambda) \subset \Omega$ for some $\lambda \in (0,1), a \in \mathbf{R}^d, |a| = 1$, it follows that

$$\int_{\Omega} |x - y|^{-\zeta} dy \geq \int_{C(x,a,\lambda)} |x - y|^{-\zeta} dy = \infty.$$

Hence the arguments for the case $x \in \Omega$ still hold in this case. Thus, we have (3.23) for every $x \in \overline{\Omega}$. Applying Dini's theorem for $\overline{\Omega}$, we obtain $\epsilon_n \to 0$ uniformly on $\overline{\Omega}$.

(c) Suppose $\zeta = \beta/(q-1) < d$. Fix $x \in \mathbf{R}^d$. For each $v \in W_n$, one has

$$1 = \int_{\Omega} v \leq \int_{\Omega} |v(y)||x - y|^{\beta/q}|x - y|^{-\beta/q} dy$$

$$\leq \left(\int_{\Omega} |x - y|^{\beta}|v(y)|^q dy \right)^{1/q} \left(\int_{\Omega} |x - y|^{\beta/(1-q)} dy \right)^{1-1/q},$$

i.e.

$$\int_{\Omega} |x - y|^{\beta}|v(y)|^q \geq \left(\int_{\Omega} |x - y|^{-\zeta} dy \right)^{1-q}.$$

Since this holds for every $v \in W_n$, we have

$$\epsilon_n(x) \geq \left(\int_{\Omega} |x - y|^{-\zeta} dy \right)^{1-q} \quad (n \in \mathbf{N})$$

and

$$\lim_{n \to \infty} \epsilon_n(x) \geq \left(\int_{\Omega} |x - y|^{-\zeta} dy \right)^{1-q} = \epsilon(x). \qquad (3.24)$$

Now for $y \neq x$, define

$$v(y) = v_x(y) = \left(\int_{\Omega} |x - \eta|^{-\zeta} d\eta \right)^{-1} |x - y|^{-\zeta}.$$

By (3.22), $v \in L^1(\Omega)$, $v > 0$ and $\int_{\Omega} v(y) dy = 1$. There exists a sequence $\{w_n\}$ in $L^q(\Omega)$ such that

$$w_n \to v \text{ in } L^1(\Omega) \text{ and a.e. in } \Omega.$$

Replacing w_n by $\min(|w_n|, v)$, we can assume that $0 \leq w_n \leq v$ in Ω.
Put

$$v_n = \left(\int_{\Omega} w_n \right)^{-1} w_n.$$

Since $v_n \in W$, we have

$$\inf_{w \in W} \int_{\Omega} |x - y|^{\beta}|w(y)|^q dy \leq \int_{\Omega} |x - y|^{\beta}|v_n(y)|^q dy$$

$$= \left(\int_{\Omega} w_n \right)^{-q} \int_{\Omega} |x - y|^{\beta}[w_n(y)]^q dy. \qquad (3.25)$$

However, for all $n \in \mathbf{N}$ and almost all $y \in \Omega$,

$$0 \leq |x - y|^{\beta}[w_n(y)]^q \leq |x - y|^{\beta}[v(y)]^q = \left(\int_{\Omega} |x - \eta|^{-\zeta} d\eta \right)^{-q} |x - y|^{-\zeta},$$

and the last function is in $L_y^1(\Omega)$. On the other hand,

$$|x - y|^{\beta}[w_n(y)]^q \to |x - y|^{\beta}[v(y)]^q \quad (n \to \infty).$$

Then, by the dominated convergence theorem,

$$\int_{\Omega} |x - y|^{\beta}[w_n(y)]^q dy \to \int_{\Omega} |x - y|^{\beta}[v(y)]^q dy = \left(\int_{\Omega} |x - y|^{-\zeta} dy \right)^{1-q}.$$

Letting $n \to \infty$ in (3.25), we obtain

$$\inf_{w \in W} \int_{\Omega} |x - y|^{\beta} |w(y)|^q dy \leq \epsilon(x).$$

Now, let $\delta > 0$ and choose $\overline{w} = \overline{w}(\delta) \in W$ such that

$$\int_{\Omega} |x - y|^{\beta} |\overline{w}(y)|^q dy \leq \inf_{w \in W} \int_{\Omega} |x - y|^{\beta} |w(y)|^q dy + \delta \leq \epsilon(x) + \delta.$$

It is easy to show that there exists a sequence $\{p_n\}$, $p_n \in W_n$, $\forall n$, such that

$$p_n \to \overline{w} \quad \text{in} \quad L^q(\Omega).$$

We have, for each $n \in N$,

$$\epsilon_n(x) \leq \int_{\Omega} |x - y|^{\beta} |p_n(y)|^q dy.$$

But

$$\int_{\Omega} |x - y|^{\beta} |p_n(y)|^q dy \to \int_{\Omega} |x - y|^{\beta} |\overline{w}(y)|^q dy \quad \text{as} \quad n \to \infty.$$

Then

$$\lim_{n \to \infty} \epsilon_n(x) \leq \lim_{n \to \infty} \int_{\Omega} |x - y|^{\beta} |p_n(y)|^q dy$$
$$= \int_{\Omega} |x - y|^{\beta} |\overline{w}(y)|^q dy$$
$$\leq \epsilon(x) + \delta.$$

Since this holds for every $\delta > 0$,

$$\lim_{n \to \infty} \epsilon_n(x) \leq \epsilon(x). \tag{3.26}$$

(3.24) and (3.26) imply

$$\lim_{n \to \infty} \epsilon_n(x) = \epsilon(x).$$

By the properties of the Fredholm integral operator (cf Chapter 1), ϵ is continuous in $\mathbf{R}^d$. Applying Dini's theorem once more, we have $\epsilon_n \to \epsilon$ uniformly on compact subsets of $\mathbf{R}^d$. Since $\epsilon(x) > 0$, $\forall x \in \mathbf{R}^d$ and $\epsilon_n \geq \epsilon$, the last conclusion of (c) follows. This completes the proof of Proposition 3.4.

Now we consider a concrete estimate for ϵ_n when the subspaces V_n approximate $L^q(\Omega)$ in the following sense :

For $m \geq 1$, there exists $C_0 = C_0(m, \Omega) > 0$ such that for all $u \in W^{m,q}(\Omega)$, there is $p_n \in V_n$ satisfying

$$\|u - p_n\|_q \leq C_0\|u\|_{m,q} n^{-m}. \tag{3.27}$$

Here, as usual, we use the notations

$$\|\cdot\|_q = \|\cdot\|_{L^q(\Omega)}, \quad \|\cdot\|_{m,q} = \|\cdot\|_{W^{m,q}(\Omega)}$$

with $m \geq 1$ and $1 \leq q \leq \infty$.

In view of Proposition 3.4, we see that $\epsilon_n \to 0$ if and only if $\beta \geq (q-1)d$. The following proposition gives the order of convergence of ϵ_n in this case.

Proposition 3.5. *Suppose (3.27) holds and that Ω satisfies a uniform cone condition.*

If $\beta > (q-1)d$, then for all

$$0 < s < \frac{\beta}{q-1} - d,$$

there exists $C = C(q, \beta, s, \Omega) > 0$ so that

$$\|\epsilon_n\|_\infty \leq Cn^{s(1-q)}.$$

If $\beta = (q-1)d$, then there exists $C = C(q, \beta, \Omega) > 0$ so that

$$\|\epsilon_n\|_\infty \leq C(\ln n)^{1-q}.$$

Proof . The proof is divided into three steps

Step 1. Put $\gamma = \beta/(q-1)$ ($\gamma \geq d$). Let $m \in \mathbf{N}$, $x \in \overline{\Omega}$ and $\delta \in (0,1)$. We show that there exists a function $\overline{v}_\delta$ with the following properties :

$$\begin{cases} \overline{v}_\delta \in C^m(\overline{\Omega}), \ \overline{v}_\delta \geq 0 \text{ on } \overline{\Omega} \\ \overline{v}_\delta(y) = |x - y|^{-\gamma} \text{ if } |x - y| \geq \delta \\ \sup_{y \in \Omega, |x-y| \leq \delta} \left\{ \max_{|p| \leq j} |D^p \overline{v}_\delta(y)| \right\} \leq C_1 \delta^{-\gamma - j}, \\ 0 \leq j \leq m, \ C_1 = C_1(m, \Omega). \end{cases} \tag{3.28}$$

Indeed, by classical extension theorems (see for example Theorem 1, §4, ch. III, [Mi]), there exists $w_1 \in C^m(\mathbf{R}^d)$ such that $w_1 \geq 0$ in $\mathbf{R}^d$ and $w_1(y) = |y|^{-\gamma}$ if $|y| \geq 1$. Put

$$C(m) = \sup_{|y| \leq 1, |q| \leq m} |D^q w_1(y)|$$

and

$$w_\delta(y) = \delta^{-\gamma} w(y/\delta) \ (0 < \delta < 1).$$

It is clear that

$$w_\delta \in C^m(\mathbf{R}^d), w_\delta \ge 0 \text{ in } \mathbf{R}^d, \text{ and } w_\delta(y) = |y|^{-\gamma} \text{ if } |y| \ge \delta.$$

For $|y| \le \delta, |p| \le j, j = 0, 1, ..., m$, we have

$$|D^p w_\delta(y)| = \delta^{-\gamma}(1/\delta)^{|p|} |D^p w_1(y/\delta)|$$
$$\le \delta^{-\gamma-j} |D^p w_1(y/\delta)| \le C(m)\delta^{-\gamma-j}.$$

Then, it can be shown that the function

$$\bar{v}_\delta(y) = w_\delta(x - y) \ \ (y \in \mathbf{R}^d)$$

satisfies (3.28).

In the following, we fix $m \in \mathbf{N}$ and define

$$\delta = \delta(n) = n^{-l} \text{ with } l = l(m) = mq[mq + \beta + (q-1)d]^{-1}. \qquad (3.29)$$

Fix $x \in \Omega$ and put

$$v_n = \left(\int_\Omega \bar{v}_\delta\right)^{-1} \bar{v}_\delta.$$

We estimate $\|v_n\|_{m,q}$.

From (3.28) and the uniform cone condition :

$$\int_\Omega \bar{v}_\delta \ge \int_{\{ y \in \Omega \ | \ |x-y| \ge \delta \}} |x - y|^{-\gamma} dy$$
$$\ge C \int_\delta^\lambda \rho^{-\gamma} \rho^{d-1} d\rho$$
$$= \begin{cases} \dfrac{C}{\gamma - d} \left[\left(\dfrac{1}{\delta}\right)^{\gamma-d} - \left(\dfrac{1}{\lambda}\right)^{\gamma-d} \right] & (\gamma > d) \\ C \ln(\lambda/\delta) & (\gamma = d). \end{cases}$$

For δ sufficiently small (i.e. n sufficiently large), for example $n \ge n_0, n_0 = n_0(\lambda, \delta, \gamma)$, we have, with $C_2 = C_2(\lambda, d, \gamma) = C_2(q, \beta, \Omega, m)$,

$$\int_\Omega \bar{v}_\delta \ge \begin{cases} C_2(1/\delta)^{\gamma-d} & (\gamma > d) \\ C_2 \ln(1/\delta) & (\gamma = d). \end{cases} \qquad (3.30)$$

By induction, we can prove that, with $C = C(m, \gamma)$,

$$|D^p(|y|^{-\gamma})| \le C|y|^{-\gamma-j}$$

for all $y \in \mathbf{R}^d \setminus \{(0, ..., 0)\}, \ \forall |p| \le j, \ j = 0, 1, ..., m.$
Then

$$|D^p(|x - y|^{-\gamma})| \le C|x - y|^{-\gamma-j} \text{ a.e. } y \in \Omega.$$

Letting $|p| = j \le m$, one has

$$\|D^p \bar{v}_\delta\|_q^q = \left(\int_{\{y\in\Omega \,:\, |x-y|\leq\delta\}} + \int_{\{y\in\Omega \,:\, |x-y|>\delta\}} \right) |D^p \bar{v}_\delta(y)|^q dy$$

$$\leq C(\frac{1}{\delta})^{(\gamma+j)q} \int_0^\delta \rho^{d-1} d\rho +$$

$$+ C \int_{\{y\in\Omega \,:\, \delta\leq|x-y|\leq R\}} |D^p(|x-y|^{-\gamma}|)^q dy \quad (R = \text{diam}\Omega)$$

$$\leq C(\frac{1}{\delta})^{q(\gamma+j)-d} + C \int_{\{y\in\Omega \,:\, \delta\leq|x-y|\leq R\}} |x-y|^{-\gamma-j} dy$$

$$\leq C(\frac{1}{\delta})^{q(\gamma+j)-d} + C \int_\delta^R \rho^{-\gamma-j} \rho^{d-1} d\rho$$

$$= C(\frac{1}{\delta})^{q(\gamma+j)-d} + C \left[(\frac{1}{\delta})^{q(\gamma+j)-d} - (\frac{1}{R})^{q(\gamma+j)-d} \right]$$

$$\leq C(\frac{1}{\delta})^{q(\gamma+j)-d}$$

with $C = C(q, \beta, \Omega, m)$, $j = 0, 1, ..., m$.

It follows that, with $C = C(q, \beta, \Omega, m)$,

$$\|\bar{v}_\delta\|_{m,q}^q = \sum_{j=0}^m \sum_{|p|=j} \int_\Omega |D^p \bar{v}_\delta(y)|^q dy \leq C(1/\delta)^{q(\gamma+m)-d}.$$

This and (3.30) imply

$$\|v_n\|_{m,q}^q = \left(\int_\Omega \bar{v}_\delta \right)^{-q} \|\bar{v}_\delta\|_{m,q}^q$$

$$\leq C_3 \begin{cases} (1/\delta)^{qm+(q-1)d} & (\gamma > d) \\ |\ln\delta|^{-q}(1/\delta)^{qm+(q-1)d} & (\gamma = d) \end{cases}$$

with $C_3 = C_3(q, \beta, \Omega, m)$.

Now from (3.27), for every n, there exists a function $\bar{p}_n \in V_n$ such that

$$\|v_n - p_n\|_q \leq C_0 \|v\|_{m,q} n^{-m}$$

$$\leq C_3^{1/q} C_0 n^{-m} \begin{cases} n^{l[qm+(q-1)d]/q} & (\gamma > d) \\ \frac{1}{l\ln n} n^{l[qm+(q-1)d]/q} & (\gamma = d) \end{cases}$$

$$\leq C_4 \begin{cases} n^{l(m+d/p)-m} & (\gamma > d) \\ (\ln n)^{-1} n^{l(m+d/p)-m} & (\gamma = d) \end{cases}$$

with $C_4 = C_4(q, \beta, \Omega, m)$.

Step 2. Let

$$p_n = \left(\int_\Omega \bar{p}_n \right)^{-1} \bar{p}_n.$$

We estimate $\|v_n - p_n\|_n$. We have

$$\left| 1 - \int_{\Omega} \overline{p}_n \right| = \left| \int_{\Omega} v_n - \int_{\Omega} \overline{p}_n \right|$$
$$\leq \|v_n - \overline{p}_n\|_1$$
$$\leq |\Omega|^{1/p} \|v_n - \overline{p}_n\|_q$$
$$\leq C_4 |\Omega|^{1/p} \begin{cases} n^{l(m+d/p)-m} & (\gamma > d) \\ (\ln n)^{-1} n^{l(m+d/p)-m} & (\gamma = d). \end{cases} \tag{3.31}$$

But from (3.29),

$$l(m + d/p) - m = -\frac{m\beta/q}{m + d/p + \beta/q} < 0.$$

Then there exists $n_1 = n_1(q, \beta, \Omega, m) \geq n_0$ such that the term in RHS of (3.31) is bounded by $1/2$ for all $n \geq n_1$. Hence in this case, $\int_{\Omega} \overline{p}_n \geq 1/2$. Then p_n is defined and $p_n \in W_n$. Moreover,

$$\|\overline{p}_n - p_n\|_q = \|\overline{p}_n\|_q \left| 1 - \int_{\Omega} \overline{p}_n \right| \left| \int_{\Omega} \overline{p}_n \right|^{-1}$$
$$\leq 2\|\overline{p}_n\|_q \left| 1 - \int_{\Omega} \overline{p}_n \right|$$
$$\leq 2C_4 |\Omega|^{1/p} \|\overline{p}_n\|_q \begin{cases} n^{l(m+d/p)-m} & (\gamma > d) \\ (\ln n)^{-1} n^{l(m+d/p)-m} & (\gamma = d) \end{cases} \tag{3.32}$$

But, from the above estimate for $\|v_n\|_{m,q}$, we have (for $m = 0$)

$$\|v_n\|_q = \left| \int_{\Omega} \overline{v}_\delta \right|^{-1} \|\overline{v}_\delta\|_q \leq C \begin{cases} n^{ld/p} & (\gamma > d) \\ (\ln n)^{-1} n^{ld/p} & (\gamma = d) \end{cases}$$

Hence

$$\|\overline{p}_n\|_q \leq \|v_n\|_q + \|v_n - \overline{p}_n\|_q$$
$$\leq C \begin{cases} n^{ld/p} + n^{l(m+d/p)-m} & (\gamma > d) \\ (\ln n)^{-1} [n^{ld/p} + n^{l(m+d/p)-m}] & (\gamma = d) \end{cases}$$
$$\leq 2C \begin{cases} n^{ld/p} & (\gamma > d) \\ (\ln n)^{-1} n^{ld/p} & (\gamma = d). \end{cases}$$

Substituting this into (3.32), we obtain, with $C_5 = C_5(q, \beta, \Omega, m)$,

$$\|\overline{p}_n - p_n\|_q \leq C_5 \begin{cases} n^{l(m+2d/p)-m} & (\gamma > d) \\ (\ln n)^{-1} n^{l(m+2d/p)-m} & (\gamma = d). \end{cases}$$

It follows that

$$\|v_n - p_n\|_q \leq \|v_n - \overline{p}_n\|_q + \|\overline{p}_n - p_n\|_q$$
$$\leq C \begin{cases} n^{l(m+d/p)-m} + n^{l(m+2d/p)-m} & (\gamma > d) \\ (\ln n)^{-1} [n^{l(m+d/p)-m} + n^{l(m+2d/p)-m}] & (\gamma = d) \end{cases}$$
$$\leq C_6 \begin{cases} n^{l(m+2d/p)-m} & (\gamma > d) \\ (\ln n)^{-1} n^{l(m+2d/p)-m} & (\gamma = d). \end{cases} \tag{3.33}$$

Now, using the arguments in [KSB], we shall obtain an estimate for

$$\int_\Omega |x-y|^\beta |v_n(y)|^q dy.$$

In fact, we have, with $C = C(q, \beta, \Omega, m)$,

$$\int_\Omega |x-y|^\beta |\bar{v}_\delta(y)|^q dy = \left(\int_{\{y\in\Omega \ : \ |x-y|\leq\delta\}\cup\{y\in\Omega \ : \ |x-y|>\delta\}} \right) |x-y|^\beta |\bar{v}_\delta(y)|^q dy$$

$$\leq C \int_0^\delta \rho^\beta \delta^{-\gamma q} \rho^{d-1} d\rho + C \int_\delta^R \rho^\beta \rho^{-\gamma q} \rho^{d-1} d\rho$$

$$= C(1/\delta)^{\gamma-d} + C \begin{cases} (1/\delta)^{\gamma-d} - (1/R)^{\gamma-d} & (\gamma > d) \\ \ln(R/\delta) & (\gamma = d). \end{cases}$$

$$\leq C \begin{cases} (1/\delta)^{\gamma-d} & (\gamma > d) \\ \ln(1/\delta) & (\gamma = d). \end{cases} \tag{3.34}$$

This and (3.30) give, with $C_7 = C_7(q, \beta, \Omega, m)$,

$$\int_\Omega |x-y|^\beta v_n^q(y) dy = \left(\int_\Omega \bar{v}_\delta \right)^{-q} \int_\Omega |x-y|^\beta [\bar{v}_\delta(y)]^q dy$$

$$\leq C \begin{cases} (1/\delta)^{(1-q)(\gamma-d)} & (\gamma > d) \\ |\ln\delta|^{1-q} & (\gamma = d). \end{cases}$$

$$\leq C_7 \begin{cases} n^{l(1-q)(\gamma-d)} & (\gamma > d) \\ (\ln n)^{q-1} & (\gamma = d). \end{cases}$$

Step 3. Since $p_n \in W_n$ $(n \geq n_1)$, we have (by the elementary inequality $(x+y)^q \leq 2^{q-1}(x^q + y^q)$ for $x, y \geq 0, q > 1$),

$$\varepsilon_n(x) \leq \int_\Omega |x-y|^\beta |p_n(y)|^q dy$$

$$\leq \int_\Omega |x-y|^\beta (|v_n(y)| + |p_n(y) - v_n(y)|)^q dy$$

$$\leq 2^{q-1} \left[\int_\Omega |x-y|^\beta |v_n(y)|^q dy + \int_\Omega |x-y|^\beta |p_n(y) - v_n(y)|^q dy \right]$$

$$\leq 2^{q-1} \left[\int_\Omega |x-y|^\beta |v_n(y)|^q dy + R^\beta \|p_n - v_n\|_q^q \right]$$

$$\leq C \begin{cases} n^{l(1-q)(\gamma-d)} + n^{l[qm+2(q-1)d]-qm} & (\gamma > d) \\ \dfrac{1}{(\ln n)^{q-1}} + \dfrac{n^{l(1-q)(\gamma-d)} + n^{l[qm+2(q-1)d]-qm}}{(\ln n)^q} & (\gamma = d), \end{cases} \tag{3.35}$$

by (3.33) and (3.34). But from (3.29), it can be seen that

$$l[m + 2(q-1)d] - qm = l(1-q)(\gamma - d) \begin{cases} < 0 \text{ if } \gamma > d \\ = 0 \text{ if } \gamma = d. \end{cases}$$

Then the term in RHS of (3.35) is bounded by

$$C_8 \begin{cases} (1/n)^{l(q-1)(\gamma-d)} & (\gamma > d) \\ (\ln n)^{1-q} & (\gamma = d) \end{cases} \tag{3.36}$$

for all $n \geq n_1$, where $C_8 = C_8(q, \beta, \Omega, m)$. Since $n_1 = n_1(q, \beta, \Omega, m)$, changing the constant C_8, we can assume that (3.36) holds for every n.

If $\gamma > d$, then from (3.29), $l(m) \to 1$ as $m \to \infty$. Hence, for $0 < s < \gamma - d$, there exists $m = m(q, \beta, \Omega, m)$ so that $[l(m)](\gamma - d) > s$. Therefore, for this m, the term in RHS of (3.36) is bounded by $C_{(}1/n)^{s(q-1)}$ with $C_8 = C_8(q, \beta, \Omega, m) = C_8(q, \beta, \Omega, s)$.

Since this is true for all $x \in \overline{\Omega}$, we obtain the first estimate of Proposition 3.5. If $\gamma = d$, we choose $m = 1$ and then $C_8 = C_8(q, \beta, \Omega)$, and (3.36) gives us the second estimate of Proposition 3.5.

3.3.2 Main results

Let us recall first some definitions about Sobolev spaces and set some notations. For $\sigma \in [0, 1), q \in (1, \infty)$,

$$u \in W^{\sigma,q}(\Omega) \Leftrightarrow \frac{|u(x) - u(y)|}{|x - y|^{\sigma + d/q}} \in L^q(\Omega \times \Omega),$$

$$\|u\|_{\sigma,q} = \left(\int_{\Omega \times \Omega} \frac{|u(x) - u(y)|^q}{|x - y|^{d + \sigma q}} dx dy \right)^{\frac{1}{q}}.$$

For $\sigma \in (0, 1), \; q = \infty$,

$$u \in W^{\sigma,\infty}(\Omega) \Leftrightarrow \frac{|u(x) - u(y)|}{|x - y|^{\sigma}} \in L^{\infty}(\Omega \times \Omega),$$

$$\|u\|_{\sigma,\infty} = \sup_{(x,y) \in \Omega \times \Omega} \frac{|u(x) - u(y)|}{|x - y|^{\sigma}}.$$

Here, $W^{\sigma,\infty}(\Omega)$ is the space of functions which are Hölder continuous in Ω with exponent σ. For a sequence $\mu = (\mu_1, \mu_2, ...)$ of real numbers, we denote by $\|\mu\|_{\infty}$ the sup norm of μ :

$$\|\mu\|_{\infty} = \sup_{j \in N} |\mu_j|.$$

For $n \in N$, we denote by $u^n(\mu)$ the BG solution corresponding to $\mu_1, ..., \mu_n$ and $g_1, ..., g_n$:

$$u^n(\mu) = \sum_{j=1}^{n} \mu_j v_j^n$$

where v_j^n, $j = 1, ..., n$, are the BG basis functions of order n, constructed in Section 3.1 (remark that $v_1^n, ..., v_n^n$ depend only on n, q, β, Ω and $g_1, ..., g_n$).

Theorem 3.6. *Suppose Ω satisfies a cone condition and let u be the solution of (MP) corresponding to the sequence*

$$\mu^0 = (\mu_1^0, \mu_2^0, ...).$$

(a) If $u_0 \in W^{\sigma,\infty}(\Omega)$, $\sigma \in (0,1)$ (i.e., u_0 is Hölder continuous with exponent σ in Ω), and

$$(q-1)d \le \beta < (q-1)d + \sigma q,$$

then for each $\delta \in (0, \delta_0)$ $(\delta_0 > 0)$, there exist $n(\delta)$ and $\eta(\delta) > 0$ such that

$$\lim_{\delta \to 0} \eta(\delta) = 0 \qquad (3.37)$$

and for every sequence μ satisfying

$$\|\mu - \mu^0\|_\infty \le \delta, \qquad (3.38)$$

we have

$$\|u^{n(\delta)}(\mu) - u_0\|_\infty \le \eta(\delta).$$

(b) If $u_0 \in W^{\sigma,\max(\gamma,p)}(\Omega)$, $\sigma \in [0,1)$, $\gamma \in [1,\infty)$, and

$$\begin{cases} (q-1)d \le \beta < (q-1)d + \sigma q \ \text{if} \ \gamma > q \\ (q-1)d \le \beta \le (q-1)d + \sigma q \ \text{if} \ \gamma \le q, \end{cases}$$

then for each $\delta \in (0, \delta_0)$ $(\delta_0 > 0)$, there exist $n(\delta) \in \mathbf{N}$ and $\eta(\delta) > 0$ such that (3.37) holds and that for every sequence μ satisfying (3.38), we have

$$\|u^{n(\delta)}(\mu) - u_0\|_\gamma \le \eta(\delta).$$

Here $n(\delta)$, $\eta(\delta)$ will be given explicitly by (3.42), (3.49), (3.53).

Proof.

(a) Observe first that for each $n \in \mathbf{N}$, the function $(x,v) \mapsto S_x^n(v)$ is continuous on $\mathbf{R}^d \times \mathbf{R}^n$. Now, since $S_x^n(v) > 0$, $\forall x \in \mathbf{R}^d$, $\forall v \in \mathbf{R}^n \setminus \{(0, ..., 0)\}$, we see that

$$C(n) = \min \left\{ S_x^n(v) \ : \ x \in \overline{\Omega}, \ |v| = \sum_{j=1}^n |v_j| = 1 \right\} > 0$$

($C(n)$ depends only on n, q, β, Ω and $\{ g_j \ : \ j \in \mathbf{N} \}$). This implies that

$$S_x^n(v) \ge C(n)|v|^q, \quad \forall v \in \mathbf{R}^n, \ \forall x \in \overline{\Omega}.$$

Now, since $\int_\Omega g_1 \ne 0$ for at least one $n \in \mathbf{N}$, we can assume without loss of generality that $\int_\Omega g_1 = 1$. Since $v_0 = (1, 0, ..., 0) \in L_n$, $n = 1, 2, ...$, we have

$$C(n)|v^n(x)|^q \le S_x^n(v^n(x)) \le S_x^n(v_0)$$

$$= \int_\Omega |x - y|^\beta |g_1(y)|^q dy \le (\text{diam } \Omega)^\beta \|g_1\|_q^q.$$

Since this holds for every $x \in \Omega$, we obtain

$$\|v^n\|_\infty \leq (C(n))^{-1/q}(\text{diam }\Omega)^{\beta/q}\|g_1\|_q \quad (n = 1, 2, ...). \tag{3.39}$$

Let us now choose a function f continuous and strictly increasing on $[1, \infty)$ to $[(C(1))^{-1/q}, \infty)$ so that

$$f(n) \geq (C(n))^{-1/q} \quad \forall n \in N. \tag{3.40}$$

We can choose, for example,

$$f(t) = t - 1 + (t - [t])(C([t] + 1))^{-1/q} + \sum_{k=1}^{[t]}(C(k))^{-1/q} \tag{3.41}$$

where $[t]$ denotes the (unique) integer such that $[t] \leq t < [t] + 1$. (The function f defined by (3.41) is equal to

$$n - 1 + \sum_{k=1}^{n}(C(k))^{-1/q}$$

at each $n \in N$ and is affine in the interval $[n, n + 1)$, $n \in N$). Now let

$$\delta_0 = (C(1))^{2/q} = \min_{x \in \Omega}\int_\Omega |x - y|^\beta|g_1(y)|^q dy.$$

For $0 < \delta \leq \delta_0$, we have $\delta^{-1/2} \geq \delta_0^{-1/2} = (C(1))^{-1/q}$, hence $f^{-1}(\delta^{-1/2}) \in [1, \infty)$. We choose

$$n(\delta) = [f^{-1}(\delta^{-1/2})] \in N. \tag{3.42}$$

Then $n(\delta) \leq f^{-1}(\delta^{-1/2})$, and by (3.39), (3.40),

$$\begin{aligned}
\|v^{n(\delta)}\|_\infty &\leq (\text{diam }\Omega)^{\beta/q}\|g_1\|_q(C(n(\delta)))^{-1/q} \\
&\leq (\text{diam }\Omega)^{\beta/q}\|g_1\|_q f(n(\delta)) \\
&\leq (\text{diam }\Omega)^{\beta/q}\|g_1\|_q \delta^{-1/2}.
\end{aligned} \tag{3.43}$$

Now, let

$$u^n(\mu^0) = \sum_{j=1}^{n}\mu_j^0 v_j^n$$

be the BG solution corresponding to $\mu_1^0, ..., \mu_n^0$ and $g_1, ..., g_n$. Then

$$\|u^n(\mu) - u_0\|_\infty \leq \|u^n(\mu) - u^n(\mu^0)\|_\infty + \|u^n(\mu^0) - u_0\|_\infty. \tag{3.44}$$

We estimate the terms in the right-hand side of (3.44). One has

$$\begin{aligned}
\|u^n(\mu) - u^n(\mu^0)\|_\infty &= \left\|\sum_{j=1}^{n}(\mu_j - \mu_j^0)v_j^n\right\|_\infty \\
&\leq \max_{1 \leq j \leq n}|\mu_j - \mu_j^0| \sup_{x \in \overline{\Omega}}\left\{\sum_{j=1}^{n}|v_j^n(x)|\right\} \\
&\leq \|\mu - \mu^0\|_\infty\|v^n\|_\infty.
\end{aligned}$$

(3.43) implies
$$\|u^{n(\delta)}(\mu) - u^{n(\delta)}(\mu^0)\|_\infty \le \delta\|v^{n(\delta)}\|_\infty$$

$$\le (\operatorname{diam} \Omega)^{\beta/q}\|g_1\|_q\delta^{1/2}. \tag{3.45}$$

We next estimate $\|u^n(\mu^0) - u_0\|_\infty$. Let $x \in \Omega$. Then we have

$$|u^n(\mu^0)(x) - u_0(x)|$$

$$= \left|\sum_{j=1}^n \mu_j^0 v_j(x) - u_0(x)\right|$$

$$= \left|\sum_{j=1}^n \left(\int_\Omega u_0(y)g_j(y)dy\right) v_j(x) - u_0(x)\int_\Omega \sum_{j=1}^n v_j(x)g_j(y)dy\right|$$

$$= \left|\int_\Omega (u_0(y) - u_0(x))\left(\sum_{j=1}^n v_j(x)g_j(y)\right) dy\right|$$

$$\le \int_\Omega |u_0(y) - u_0(x)|\left|\sum_{j=1}^n v_j(x)g_j(y)\right| dy \tag{3.46}$$

$$\le \|u_0\|_{\sigma,\infty} \int_\Omega |x - y|^{\sigma-\beta/q}\left(|x-y|^{\beta/q}\left|\sum_{j=1}^n v_j(x)g_j(y)\right|\right) dy$$

$$\le \|u_0\|_{\sigma,\infty} \left(\int_\Omega |x - y|^{(\sigma-\beta/q)p}dy\right)^{1/p} \times$$

$$\times \left(\int_\Omega |x - y|^\beta\left|\sum_{j=1}^n v_j(x)g_j(y)\right|^q dy\right)^{1/q}. \tag{3.47}$$

But by hypothesis,
$$-\left(\sigma - \frac{\beta}{q}\right)\frac{q}{q-1} = \frac{\beta - \sigma q}{q-1} < d.$$

Hence the function
$$x \mapsto \int_\Omega |x - y|^{(\sigma-\beta/q)p}dy$$

is continuous on $\mathbf{R}^d$, and, therefore

$$C \equiv C(q, \beta, \Omega, \sigma) \equiv \sup_{x\in\Omega}\left(\int_\Omega |x - y|^{(\sigma-\beta/q)p}dy\right)^{1/p} < \infty.$$

It follows that the right-hand side of (3.47) is bounded by $C\|u_0\|_{\sigma,\infty}[\epsilon_n(x)]^{1/q}$. We then obtain
$$|u^n(\mu^0)(x) - u_0(x)| \le C\|u_0\|_{\sigma,\infty}[\epsilon_n(x)]^{1/q}.$$

Since this holds for every $x \in \Omega$, we have

$$\|u^n(\mu^0) - u_0\|_\infty \le C\|u_0\|_{\sigma,\infty}\|\epsilon_n\|_\infty^{1/q} \tag{3.48}$$

with $C = C(q, \beta, \Omega, \sigma)$. From (3.44), (3.45), (3.48), we see that $\|u^{n(\delta)}(\mu) - u_0\|_\infty \le \eta(\delta)$ with

$$\eta(\delta) = (\text{diam } \Omega)^{\beta/q}\|g_1\|_q \delta^{1/2} + C(q, \beta, \Omega, \sigma)\|u_0\|_{\sigma,\infty}\|\epsilon_{n(\delta)}\|_\infty^{1/q}. \tag{3.49}$$

Now, as $\delta \to 0$, we have $\delta^{-1/2} \to \infty$. Hence, $f^{-1}(\delta^{-1/2}) \to \infty$ and $n(\delta) = [f^{-1}(\delta^{-1/2})] \to \infty$. Therefore $\|\epsilon_{n(\delta)}\|_\infty \to 0$ by Proposition 3.4(b). This means that $\eta(\delta) \to 0$ as $\delta \to 0$ and completes the proof of (a).

(b) Let $u^n(\mu^0)$ be as in the proof of (a). By Proposition 3.2, $u^n(\mu^0) \in C(\overline{\Omega}) \subset L^\gamma(\Omega)$, and as in (3.44), we have

$$\|u^n(\mu) - u_0\|_\gamma \le \|u^n(\mu) - u^n(\mu^0)\|_\gamma + \|u^n(\mu^0) - u_0\|_\gamma. \tag{3.50}$$

Choose δ_0 and $n(\delta)$ as in (a). In view of (3.45), we obtain

$$\|u^{n(\delta)}(\mu) - u^{n(\delta)}(\mu^0)\|_\gamma \le |\Omega|^{1/\gamma}\|u^{n(\delta)}(\mu) - u^{n(\delta)}(\mu^0)\|_\infty$$
$$\le |\Omega|^{1/\gamma}(\text{diam } \Omega)^{\beta/q}\|g_1\|_q\delta^{1/2}. \tag{3.51}$$

Now, we estimate $\|u^n(\mu^0) - u_0\|_\gamma$. We shall show that

$$\|u^n(\mu^0) - u_0\|_\gamma \le C\|\epsilon_n\|_\infty^{1/q}\|u_0\|_{\sigma,\max(\gamma,p)} \tag{3.52}$$

where $C = C(q, \beta, \gamma, \Omega, \sigma)$. We consider three cases.

Case (i): $p < \gamma < \infty$, (ii) $\gamma = p$, (iii) $1 \le \gamma < p$.

In case (i), $\max(\gamma, p) = \gamma$ and there exists $1 < p < \infty$ such that $1/p + 1/\gamma = 1/p$. Put $\lambda = d/\gamma + \sigma$. Then by hypothesis,

$$-(\lambda - \beta/q)p = (\beta/q - \lambda)p < (d/p + \sigma - d/\gamma - \sigma)p = dp(1/p - 1/\gamma) = d$$

Therefore, the function

$$x \mapsto \int_\Omega |x - y|^{(\lambda - \beta/q)p} dy$$

is continuous on $\mathbf{R}^d$, and

$$C \equiv C(q, \beta, \gamma, \Omega, \sigma) \equiv \sup_{x \in \Omega} \left(\int_\Omega |x - y|^{(\lambda - \beta/q)p} dy\right)^{1/p} < \infty$$

For $x \in \Omega$, Hölder's inequality gives us

$$\int_\Omega |u_0(x) - u_0(y)| \left|\sum_{j=1}^n v_j(x)g_j(y)\right| dy$$

$$= \int_\Omega |x - y|^{\lambda - \beta/q} \left(|x - y|^{\beta/q}\left|\sum_{j=1}^n v(x)g_j(y)\right|\right) \frac{|u_0(x) - u_0(y)|}{|x - y|^\lambda}$$

$$\le \left(\int_\Omega |x - y|^{(\lambda - \beta/q)p} dy\right)^{1/p} \left(\int_\Omega |x - y|^\beta \left|\sum_{j=1}^n v_j(x)g_j(y)\right|^q dy\right)^{1/q} \times$$

$$\times \left(\int_\Omega \frac{|u_0(x) - u_0(y)|^\gamma}{|x - y|^{\lambda\gamma}} dy\right)^{1/\gamma}.$$

This and (3.46) imply

$$
\begin{aligned}
\|u^n(\mu^0) - u_0\|_\gamma &= \left(\int_\Omega |u^n(\mu^0)(x) - u_0(x)|^\gamma dx \right)^{1/\gamma} \\
&\leq \left\{ \int_\Omega \left(\int_\Omega |x-y|^{(\lambda-\beta/q)p} dy \right)^{\gamma/p} (\epsilon_n(x))^{\gamma/q} \times \right. \\
&\qquad \left. \left(\int_\Omega \frac{|u_0(x)-u_0(y)|^\gamma}{|x-y|^{\lambda\gamma}} dy \right) dx \right\}^{1/\gamma} \\
&\leq \left\{ \int_\Omega C^\gamma(\epsilon_n(x))^{\gamma/q} \left(\int_\Omega \frac{|u_0(x)-u_0(y)|^\gamma}{|x-y|^{\lambda\gamma}} dy \right) dx \right\}^{1/\gamma} \\
&\leq C \|\epsilon_n\|_\infty^{1/q} \left(\int_{\Omega\times\Omega} \frac{|u_0(x)-u_0(y)|^\gamma}{|x-y|^{d+\sigma\gamma}} dy dx \right)^{1/\gamma}.
\end{aligned}
$$

This is (3.52).

Case (ii) : $\gamma = p$. Put $\lambda = d/p + \sigma$. Then $\lambda - \beta/q \geq 0$ and the function $(x,y) \mapsto |x-y|^{\lambda-\beta/q}$ is continuous on $\mathbf{R}^d \times \mathbf{R}^d$. Hence

$$
C \equiv C(q,\beta,\Omega,\sigma) \equiv \sup_{(x,y)\in\Omega\times\Omega} |x-y|^{\lambda-\beta/q} < \infty.
$$

In this case,

$$
\begin{aligned}
\int_\Omega |u_0(x) - u_0(y)| &\left| \sum_{j=1}^n v_j(x) g_j(y) \right| dy = \\
&= \int_\Omega |x-y|^{\lambda-\beta/q} \left(|x-y|^{\beta/q} \left| \sum_{j=1}^n v_j(x) g_j(y) \right| \right) \frac{|u_0(x)-u_0(y)|}{|x-y|^\lambda} dy \\
&\leq C \int_\Omega \left(|x-y|^{\beta/q} \left| \sum_{j=1}^n v_j(x) g_j(y) \right| \right) \frac{|u_0(x)-u_0(y)|}{|x-y|^\lambda} dy \\
&\leq C \left(\int_\Omega |x-y|^\beta \left| \sum_{j=1}^n v_j(x) g_j(y) \right|^q dy \right)^{1/q} \left(\int_\Omega \frac{|u_0(x)-u_0(y)|^p}{|x-y|^{\lambda p}} dy \right)^{1/p} \\
&= C(\epsilon_n(x))^{1/q} \left(\int_\Omega \frac{|u_0(x)-u_0(y)|^{q\star}}{|x-y|^{\lambda p}} dy \right)^{1/p}.
\end{aligned}
$$

This and (3.46) imply

$$
\begin{aligned}
\|u^n(\mu^0) - u_0\|_p &= \left(\int_\Omega |u^n(\mu^0)(x) - u_0(x)|^p dx \right)^{1/p} \leq \\
&\leq C \|\epsilon_n\|_\infty^{1/q} \left(\int_{\Omega\times\Omega} \frac{|u_0(x)-u_0(y)|^p}{|x-y|^{d+\sigma p}} dy dx \right)^{1/p}.
\end{aligned}
$$

We obtain (3.52).

Case (iii) : $1 \leq \gamma < p$. We have $\max(\gamma, p) = p$ and, by Hölder's inequality,

$$\int_{\Omega} |u^n(\mu^0)(x) - u_0(x)|^\gamma dx \leq |\Omega|^{1-\gamma/p} \left(\int_{\Omega} |u^n(\mu^0)(x) - u_0(x)|^p dx \right)^{\gamma/p}.$$

This means

$$\|u^n(\mu^0) - u_0\|_\gamma \leq |\Omega|^{1/\gamma - 1/p} \|u^n(\mu^0) - u_0\|_p.$$

Applying (ii), we obtain (3.52) in this case.

We have just proved that (3.52) holds for all cases. It follows from (3.50), (3.52), (3.52) that

$$\|u^{n(\delta)}(\mu) - u_0\|_\gamma \leq \eta(\delta)$$

with

$$\eta(\delta) = |\Omega|^{1/\gamma} (\text{diam } \Omega)^{\beta/q} \|g_1\|_q \delta^{1/2} +$$
$$+ C(q, \beta, \gamma, \Omega, \sigma) \|u_0\|_{\sigma, \max(\gamma, p)} \|\epsilon_{n(\delta)}\|_\infty^{1/q}. \tag{3.53}$$

As in (a), we have $n(\delta) \to \infty$ as $\delta \to 0$. Since $\beta \geq (q-1)d$, Proposition 3.4 (b) implies that $\|\epsilon_{n(\delta)}\|_\infty \to 0$. Hence $\eta(\delta) \to 0$ as $\delta \to 0$ and our proof is complete.

Combining Theorems 3.6 with Proposition 3.5, we obtain the following result which gives concrete estimates for $u^{n(\delta)}(\mu) - u_0$.

Theorem 3.7. *Suppose (3.27) holds and that Ω satisfies a uniform cone condition. Let u_0 be the solution of (MP) corresponding to the sequence $\mu^0 = (\mu_1^0, \mu_2^0, ...)$. Let*

$$\delta_0 = \left(\min_{x \in \overline{\Omega}} \int_{\Omega} |x - y|^\beta |g_1(y)|^q dy \right)^{2/q}$$

and choose, for $0 < \delta < \delta_0$,

$$n(\delta) = [f^{-1}(\delta^{-1/2})] \in \mathbf{N}$$

(f is given by (3.40) or (3.41)).

(a) If the conditions of Theorem 3.6(a) hold, then for any sequence μ satisfying (3.38), we have

$$\|u^{n(\delta)}(\mu) - u_0\|_\infty \leq C\delta^{1/2} +$$
$$+ C\|u_0\|_{\sigma,\infty} \cdot \begin{cases} [f^{-1}(\delta^{-1/2})]^{-s/p} & (\beta > (q-1)d) \\ (\ln[f^{-1}(\delta^{-1/2})])^{-1/p} & (\beta = (q-1)d) \end{cases}$$

where $0 < s < \beta/(q-1) - d$ and $C = C(q, \beta, \sigma, s, \Omega, \|g_1\|_q)$.

(b) If the conditions of Theorem 3.6(b) hold, then for any sequence μ satisfying (3.38), we have

$$\|u^{n(\delta)}(\mu) - u_0\|_\gamma \leq C\delta^{1/2} +$$
$$+ C\|u_0\|_{\sigma,\max(\gamma,p)} \cdot \begin{cases} [f^{-1}(\delta^{-1/2})]^{-s/p} & (\beta > (q-1)d) \\ (\ln[f^{-1}(\delta^{-1/2})])^{-1/p} & (\beta = (q-1)d) \end{cases}$$

where $0 < s < \beta/(q-1) - d$ and $C = C(q, \beta, \gamma, \sigma, s, \Omega, \|g_1\|_q)$.

Proof.

In view of Proposition 3.5 and of the choice of $n(\delta)$, we have

$$\|\epsilon_{n(\delta)}\|_\infty^{1/q} \leq C \begin{cases} [f^{-1}(\delta^{-1/2})]^{-s/p} & (\beta > (q-1)d) \\ (\ln[f^{-1}(\delta^{-1/2})])^{-1/p} & (\beta = (q-1)d) \end{cases}$$

where $C = C(q, \beta, s, \Omega)$ and $0 < s < \beta/(q-1) - d$. Substituting this into (3.49) and (3.53), we obtain the estimate in Theorem 3.7.

Theorems 3.6 and 3.7 above give stability estimates for the BG solutions in case (MP) may not have a solution. In the case (MP) does have a solution, we actually have convergence to the solution and moreover sharper estimates are available. More precisely, we have the following two theorems, whose proofs are essentially contained in the proofs of Theorems 3.6 and 3.7.

Theorem 3.8. *Let u be the solution of (MP) corresponding to $\mu_1, \mu_2, \ldots$ and let u^n be the BG solution corresponding to $\mu_1, \ldots, \mu_n$ and $g_1, \ldots, g_n$ ($n = 1, 2, \ldots$).*

(a) Suppose u is Hölder continuous with exponent $\sigma \in (0,1]$ in Ω and $(q-1)d \leq \beta < (q-1)d + \sigma q$. Then $u^n \to u$ uniformly on compact subsets of Ω, and for all $x \in \Omega$,

$$|u^n(x) - u(x)| \leq C[\epsilon_n(x)]^{1/q}\|u\|_{\sigma,\infty}.$$

The convergence is uniform on $\overline{\Omega}$ if Ω satisfies a cone condition (here $C = C(q, \beta, \Omega, \sigma)$). If we suppose further that (3.27) holds and that Ω satisfies a uniform cone condition, then

$$\|u^n - u\|_\infty \leq C\|u\|_{\sigma,\infty} \begin{cases} n^{-s/p} & (\beta > (q-1)d) \\ (\ln n)^{-1/p} & (\beta = (q-1)d) \end{cases}$$

where $C = C(q, \beta, \sigma, s, \Omega)$ and $0 < s < \beta/(q-1) - d$.

(b) Suppose

$$u \in W^{\sigma,\max(\gamma,p)}(\Omega), \quad \sigma \in [0,1), \quad \gamma \in [1,\infty),$$

and

$$(q-1)d \leq \beta < (q-1)d + \sigma q \text{ if } \gamma > q,$$
$$(q-1)d \leq \beta \leq (q-1)d + \sigma q \text{ if } \gamma \leq q$$

and that Ω satisfies a cone condition. Then $u^n \to u$ in $L^\gamma(\Omega)$ and we have the estimate

$$\|u^n - u\|_\gamma \leq C\|u\|_{\sigma,\max(\gamma,p)}\|\epsilon\|_\infty^{1/q}$$

where $C = C(q, \beta, \gamma, \Omega, \sigma)$. If we suppose further that (3.27) holds and that Ω satisfies a uniform cone condition, then

$$\|u^n - u\|_\gamma \leq C\|u\|_{\sigma,\max(\gamma,p)} \begin{cases} n^{-s/p} & (\beta > (q-1)d) \\ (\ln n)^{-1/p} & (\beta = (q-1)d) \end{cases}$$

where $C = C(q, \beta, \gamma, \sigma, s, \Omega)$ and $0 < s < \beta/(q-1) - d$.

It is also interesting to consider the case when $q = 2$ and $\beta = 2k$. In this particular case, Theorem 3.8 becomes

Theorem 3.9. *Consider the case when $q = 2$ and $\beta = 2k$, $k \in N$, and let u, u^n be as in Theorem 3.8.*

(a) Suppose u is Hölder continuous with exponent $\sigma \in (0, 1]$ in Ω and choose

$$k = \begin{cases} d/2 & \text{if } d \text{ is even} \\ (d+1)/2 & \text{if } d \text{ is odd and } 1/2 < \sigma \leq 1. \end{cases}$$

Then $u^n \to u$ uniformly on compact subsets of Ω and there exists $C = C(\Omega, \sigma) > 0$ such that, for all $n \in \mathbf{N}, x \in \Omega$,

$$|u^n(x) - u(x)| \leq C\|u\|_{\sigma,\infty}[\epsilon_n(x)]^{1/2}.$$

The convergence is uniform on $\overline{\Omega}$ if Ω satisfies a cone condition. If we suppose further that

$$\Omega \text{ satisfies a uniform cone condition}$$

$$\text{and } (3.27) \text{ holds for } q = 2, \tag{3.54}$$

then, with $C = C(\sigma, s, \Omega)$,

$$\|u^n - u\|_\infty \leq C\|u\|_{\sigma,\infty} \begin{cases} n^{-s/2} & \text{if } s \in (0, 1), d \text{ is odd and } 1/2 < \sigma \leq 1 \\ (\ln n)^{-1/2} & \text{if } d \text{ is even.} \end{cases}$$

In particular, these results hold if u is Lipschitz continuous in Ω and $k = [(d+1)/2]$.
(b) Suppose $u \in W^{\sigma,\max(2,\gamma)}(\Omega)$, $\sigma \in [0, 1)$, $\gamma \in [1, \infty)$, and that Ω satisfies a cone condition. Choose

$$k = \begin{cases} d/2 \text{ if } d \text{ is even} \\ (d+1)/2 \text{ if } d \text{ is odd and } 1/2 < \sigma \leq 1, \text{ or } 1/2 \leq \sigma < 1 \text{ if } \gamma \leq 2. \end{cases}$$

Then $u^n \to u$ in $L^\gamma(\Omega)$ and there exists $C = C(\Omega, \sigma, \gamma)$ such that

$$\|u^n - u\|_\gamma \leq C\|u\|_{\sigma,\max(2,\gamma)}\|\epsilon_n\|_\infty^{1/2} \quad \forall n.$$

If we suppose further that (3.54) holds, then, with $C = C(\gamma, \sigma, s, \Omega)$,

$$\|u^n - u\|_\gamma \leq C\|u\|_{\sigma,\max(\gamma,2)}(\ln n)^{-1/2}$$

if d is even,

$$\|u^n - u\|_\gamma \leq C\|u\|_{\sigma,\max(\gamma,2)}n^{-s/2}$$

if $s \in (0, 1)$, d is odd and $1/2 < \sigma < 1$ (or $1/2 \leq \sigma < 1$ if $\gamma \leq 2$).

The foregoing Theorems 3.8 and 3.9 extend Theorem 3.1 and Theorem 4.2 of [KSB] where the solution (denoted by u) is either Lipschitzian or in $H^\alpha(\Omega)$ for $0 < \alpha < 1$.

4 The Hausdorff moment problem: regularization and error estimates

The Hausdorff moment problem has its origin in Mechanics. The problem consists in finding the distribution of (positive) mass on an interval $(a, b) \subset (-\infty, \infty)$, given the moments of order k $(k = 0, 1, 2, ...)$ of the mass distribution. If $u = u(x)$ is the mass distribution on (a, b), then the total mass on the whole interval is given by

$$\int_a^b u(x)dx,$$

while $\int_a^b xu(x)dx$ represents the static moment of the mass distribution and $\int_a^b x^2 u(x)dx$ is the moment of inertia with respect to the point $x = 0$. Stieltjes gave the name of generalized moment of order k to the integral $\int_a^b x^k u(x)dx$ and studied the problem in the case $(a, b) = (0, \infty)$. The moment problem on $(-\infty, \infty)$ was considered by Hamburger in 1920, while the Hausdorff moment problem, which is concerned with the problem on $(a, b) = (0, 1)$, was first investigated by Hausdorff in 1923.

The classical moment problems referred to above have connections with several topics in both pure and applied mathematics, such as in spectral representation of operators, partial fractions, theory of harmonic functions on a half plane, inverse problems .... Moment problems constitute a typical and very important class of ill-posed problems.

The ill-posedness of the Hausdorff moment problem and some of the regularization procedures will be studied in detail in the sequel. We shall consider the problem in the multidimensional case. Thus, let $I = (0, 1)^d \subset \mathbf{R}^d$ $(d \in \mathbf{N})$. Consider the problem of finding u in $L^2(I)$ satisfying the sequence of equations

$$\int_I u(x_1, .., x_d)x_1^{k_1}..x_d^{k_d}dx_1..dx_d = \mu_{k_1..k_d},$$

$$k_i = 0, 1, 2..., \quad i = 1, .., d \qquad (4.1)$$

where $\mu = (\mu_{k_1..k_d})$ is a given bounded sequence. When $d = 1$ the problem (4.1) is the classical Hausdorff moment problem. It can be shown that (4.1) is ill-posed, i.e., solutions do not always exist, and in the case of existence of solutions, these do not depend continuously on the given data (which, in this case, are represented by the right hand side μ). It is the purpose of this chapter to give a *regularization* of the problem by *finite moments*. The chapter is divided into two sections. Sect. 4.1 deals with the finite moment approximation of (4.1). Here, while following the general approach of [AGT], the treatment differs from the work of Chapter

2 in the crucial choice of the function $f(t)$, $[f^{-1}(\epsilon^{-1/2})]$ being the dimension of the finite moment approximation. The exposition is here based on [AGT]. Sect. 4.2 deals with the moment problem associated with the Laplace transform. Under the heading "Notes and remarks" at the end of the chapter, we shall point to a sharpening of the results in Theorem 4.2. Under the same heading, we shall further discuss the ill-posedness of the Hausdorff moment problem and the regularization of the problem in an L^p-setting, $1 < p < \infty$, using the Backus-Gilbert theory.

4.1 Finite moment approximation of (4.1)

The regularization method of moment problems in Chapter 2, we recall, is based upon the Gram-Schmidt orthonormalization process. We shall regularize (4.1) by a sequence of finite moment problems

$$\int_I u(x_1, .., x_d) x_1^{k_1} .. x_d^{k_d} dx_1 dx_2 .. dx_d = \mu_{k_1 .. k_d},$$

$$k_i = 0, 1, 2, ..., n, \quad i = 1, .., d. \tag{4.2}$$

In our construction of finite dimensional approximations, we obtain an orthonormalization of the basis functions $\{x_1^{k_1} .. x_d^{k_d}\}$, $k_i = 0, 1, .., \; i = 1, .., d$, not through the orthonormalization process but by using products of one-dimensional Gram-Schmidt orthonormalized polynomials.

Hence, a first step in our analysis is to orthonormalize the family $(1, x, x^2, ...)$. The orthonormalization will be given in terms of Legendre polynomials. Let $P_n(t)$ be the Legendre polynomial of degree n:

$$P_n(t) = \sum_{k=0}^{n} \frac{(n+k)!}{(n-k)!(k!)^2} \frac{(t-1)^k}{2^k} \tag{4.3}$$

or equivalently (cf [Co3], p. 163)

$$P_n(t) = \frac{1}{2^n n!} \frac{d^n}{dt^n} (t^2 - 1)^n. \tag{4.4}$$

By direct computations from (4.3)-(4.4), we have

$$\int_{-1}^{1} P_n(t) P_m(t) dt = 0, \quad n \neq m, \tag{4.5}$$

$$\int_{-1}^{1} P_n^2(t) dt = \frac{2}{2n+1}. \tag{4.6}$$

Define

$$L_n(x) = \sqrt{2n+1} P_n(1 - 2x). \tag{4.7}$$

Then we have by (4.5), (4.6)

$$\int_0^1 L_n(t)L_k(t)dt = 0, \quad n \neq k, \tag{4.8}$$

$$\int_0^1 L_n^2(t)dt = 1. \tag{4.9}$$

Since L_n is a polynomial of degree n, we have by (4.7)-(4.9) that (L_n) is a complete orthonormal sequence in $L^2(0,1)$. Note that (L_n) is the Gram-Schmidt orthogonalization of $(1, x, x^2, ..)$ and hence is complete in $L^2(0,1)$.

Substituting $t = 1 - 2x$ into (4.3), we get:

$$L_n(x) = \sum_{k=0}^{n} C_{nk}x^k \tag{4.10}$$

where

$$C_{nk} = (2n+1)^{1/2}(-1)^k \frac{(n+k)!}{(n-k)!(k!)^2}. \tag{4.11}$$

Now put

$$L_{k_1..k_d}(x_1,..,x_d) = L_{k_1}(x_1)...L_{k_d}(x_d). \tag{4.12}$$

Then by the completeness of $(L_n)_{n\geq 0}$ in $L^2(0,1)$, the sequence $\{L_{k_1..k_d}\}$ forms a complete orthonormal set in $L^2(I)$. In view of (4.10), (4.12), we have

$$L_{k_1..k_d}(x_1,..,x_d) = \sum_{p_1=0}^{k_1} ... \sum_{p_d=0}^{k_d} C_{k_1 p_1}..C_{k_d p_d} x_1^{p_1}..x_d^{p_d} \tag{4.13}$$

If $\mu = (\mu_{k_1..k_d})$ is a real sequence, we define the sequence

$$\lambda = \lambda(\mu) = (\lambda_{k_1..k_d}), \quad k_1, \ldots, k_d = 0, 1, \ldots,$$

as follows

$$\lambda_{k_1..k_d} = \lambda_{k_1..k_d}(\mu) = \sum_{p_1=0}^{k_1} ... \sum_{p_d=0}^{k_d} C_{k_1 p_1}..C_{k_d p_d}\mu_{p_1...p_d}.$$

Now, put

$$p^n = p^n(\mu) = \sum_{k_1,..,k_d=0}^{n} \lambda_{k_1..k_d}(\mu)L_{k_1..k_d}. \tag{4.14}$$

Then p^n is a minimal norm solution of (4.3).

The main results of this chapter are Theorem 4.1 and Theorem 4.2 below.

Theorem 4.1. *Let $\mu = (\mu_{k_1..k_d})$ be a given sequence of real numbers. Then a necessary and sufficient condition for (4.1) to have a solution is that*

$$\sum_{k_1,..,k_d=0}^{\infty} \left(\sum_{p_1,..,p_d=0}^{\infty} C_{k_1 p_1}..C_{k_d p_d}\mu_{p_1..p_d} \right)^2 < \infty \tag{4.15}$$

where C_{ij}, $i, j = 0, ..$ is defined by (4.11) if $j \leq i$ and $C_{ij} = 0$ if $j > i$.
If u is the (unique) solution of (4.1) then

$$p^n(\mu) \to u \quad in \ L^2(I) \ as \ n \to \infty. \tag{4.16}$$

Moreover, if the solution u is in $H^1(I)$ then

$$\|p^n(\mu) - u\| \leq \frac{1}{2(n+1)} \|u\|_{H^1}, \quad n = 1, 2, \ldots. \tag{4.17}$$

Remark 4.1. If the solution u is in $H^2(I)$, then it can be shown that

$$\|p^n(\mu) - u\| \leq \frac{1}{2\sqrt{2}} \|u\|_{H^2}(n+1)^{-2}. \tag{4.18}$$

Note that this estimate is probably not optimal. The derivation of (4.18) involves lengthy calculations not shown here. If u is in $H^m(I)$ then one would have a similar estimate, through some elaborate estimates.

Theorem 4.2. *Let $u_0 \in L^2(I)$ be the solution of (4.1) corresponding to $\mu^0 = (\mu^0_{k_1..k_d})$ in the right hand side. Let*

$$f(t) = \left(\frac{27}{8\sqrt{5}} (2t+1)^{1/2} 3^{2t} \right)^d, \tag{4.19}$$

and for $0 < \epsilon < 1$, put

$$n(\epsilon) = \left[f^{-1}(\epsilon^{-1/2}) \right] \tag{4.20}$$

where $[x]$ is the largest integer $\leq x$.
    Then there exists a function $\eta(\epsilon)$, $0 < \epsilon < 1$, such that $\eta(\epsilon) \to 0$ as $\epsilon \to 0$ and that for all sequences μ satisfying

$$\|\mu - \mu^0\|_\infty \equiv \sup_{k_1..k_d} |\mu_{k_1..k_d} - \mu^0_{k_1..k_d}| < \epsilon \tag{4.21}$$

we have

$$\|p^{n(\epsilon)}(\mu) - u_0\| \leq \eta(\epsilon).$$

Moreover, if $u_0 \in H^1(I)$, then

$$\|p^{n(\epsilon)}(\mu) - u_0\| \leq \epsilon^{1/2} + \frac{\|u_0\|_{H^1(I)}}{C(\epsilon)} \tag{4.22}$$

with

$$C(\epsilon) = 2(2 \ln 3 + \frac{1}{2} \ln 2)^{-1} \ln \left(\frac{8\sqrt{5}}{27\sqrt{2}\sqrt[2d]{\epsilon}} \right)$$

Remark 4.2. If u_0 is in $H^2(I)$, then using Remark 4.1, one can take the right hand side of (4.22) as

$$\epsilon^{1/2} + \frac{\|u_0\|_{H^2}}{2\sqrt{2}\tilde{C}(\epsilon)}$$

where

$$\tilde{C}^1(\epsilon) = (2\ln 3 + \frac{1}{2}\ln 2)^{-2}\ln^2\left(\frac{8\sqrt{5}}{27\sqrt{2}\,\sqrt[2d]{\epsilon}}\right).$$

Before giving the proofs of Theorem 4.1 and Theorem 4.2, we first state (and prove) the following Lemma 4.1 which is crucial for the proof of these theorems. The idea of the following simplified proof of Lemma 4.1 was kindly suggested to us by one of the referees.

Lemma 4.1. *For every v in $H^1(0,1)$, we have*

$$\sum_{k=1}^{\infty} 4k^2|\alpha_k|^2 \leq \int_0^1 |v'(x)|^2\,dx \tag{4.23}$$

where

$$\alpha_k = (v, L_k) \text{ and } (.,.) \text{ is the } L^2(0,1) - \text{inner product.}$$

Proof

We rely on the differential equation for shifted Legendre polynomials (see, e.g., [AS])

$$[x(1-x)L_k'(x)]' = k(k+1)L_k(x), \quad k = 0,1,2,\dots.$$

We first consider the case $v \in C^2[0,1]$. From the foregoing equations, we can calculate the Fourier-Legendre coefficients of $[x(1-x)v']'$ as

$$([x(1-x)v']', L_k) = k(k+1)\alpha_k, \quad k = 0,1,2,\dots,$$

where $\alpha_k = (v, L_k)$.

Using the Parseval theorem for two functions v and $[x(1-x)v']'$, the latter equialities imply

$$\int_0^1 [x(1-x)v'(x)]'\bar{v}(x)dx = \sum_{k=0}^{\infty} k(k+1)\alpha_k\bar{\alpha}_k.$$

It follows that

$$\int_0^1 x(1-x)|v'(x)|^2\,dx = \sum_{k=0}^{\infty} k(k+1)|\alpha_k|^2. \tag{4.24}$$

Since $C^2[0,1]$ is dense in $H^1(0,1)$, the latter equality also holds for every $v \in H^1(0,1)$.

Now, one has

$$x(x-1) \leq 1/4 \text{ for all } x \in [0,1].$$

Hence, (4.24) gives (4.23). This completes the proof of Lemma 4.1.

We now give the proofs of Theorem 4.1 and Theorem 4.2.

4.1.1 Proof of Theorem 4.1.

We easily get the first result of Theorem 4.1 using the completeness and orthonormality of the family $(L_{k_1..k_d})$ and considering the series representation

$$u = \sum_{k_1,..k_d=0}^{\infty} \lambda_{k_1..k_d}(\mu) L_{k_1..k_d} \tag{4.25}$$

for every u in $L^2(I)$. From (4.14) and the latter relation, we shall get (4.15), (4.16). For the proof of (4.17), we rely on Lemma 4.1. From that lemma, one has

$$\sum_{k_1,...,k_d=0}^{\infty} 4k_j^2 \lambda_{k_1..k_d}^2 \leq \int_I \left| \frac{\partial u}{\partial x_j} \right|^2 dx_1..dx_d \tag{4.26}$$

for $j = 1, 2, .., d$.

Letting $j = 1, .., d$ in (4.26) and adding together the results thus obtained, we get

$$\sum_{k_1,..,k_d=0}^{\infty} \left(4 \sum_{j=1}^{d} k_j^2 \right) \lambda_{k_1..k_d}^2 \leq \|u\|_{H^1(I)}^2. \tag{4.27}$$

Now, subtracting (4.14) from (4.25), we have

$$u - p^n(\mu) = \sum_{\max_{1 \leq i \leq d} |k_i| > n} \lambda_{k_1..k_d} L_{k_1..k_d}$$

Hence

$$\|p^n(\mu) - u\|^2 = \sum_{\max_{1 \leq i \leq d} |k_i| > n} \lambda_{k_1..k_d}^2. \tag{4.28}$$

Note that

$$4 \sum_{j=1}^{d} k_j^2 \geq 4(n+1)^2 \qquad \text{as } \max_{1 \leq i \leq d} |k_i| > n. \tag{4.29}$$

Combining (4.27)-(4.29) gives

$$(n+1)^2 \|p^n(\mu) - u\|^2 \leq \sum_{\max_{1 \leq i \leq d} |k_i| > n} (n+1)^2 \lambda_{k_1..k_d}^2$$

$$\leq \frac{1}{4} \|u\|_{H^1(I)}^2$$

which is the desired inequality. This completes the proof of Theorem 4.1.

4.1.2 Proof of Theorem 4.2.

We have

$$\|p^n(\mu) - u_0\| \leq \|p^n(\mu) - p^n(\mu^0)\| + \|p^n(\mu^0) - u_0\|. \tag{4.30}$$

Using (4.14), we have

$$p^n(\mu) - p^n(\mu^0) = \sum_{k_1,..,k_d=0}^{\infty} \sum_{p_1,...,p_d=0}^{\infty} C_{k_1 p_1}..C_{k_d p_d}(\mu_{p_1..p_d} - \mu^0_{p_1..p_d})L_{p_1..p_d}.$$

Here we recall that $C_{ij} = 0$ if $i < j$.

Hence:

$$\|p^n(\mu) - p^n(\mu^0)\|^2 = \sum_{k_1,..,k_d=0}^{\infty} \left(\sum_{p_1,...,p_d=0}^{\infty} C_{k_1 p_1}..C_{k_d p_d}(\mu_{p_1..p_d} - \mu^0_{p_1..p_d}) \right)^2$$

$$\leq \epsilon^2 \sum_{k_1...,k_d=0}^{\infty} \left(\sum_{p_1,...,p_d=0}^{\infty} |C_{k_1 p_1}..C_{k_d p_d}| \right)^2. \tag{4.31}$$

From (4.11) one has

$$C_{mj} = (2m + 1)^{1/2}(-1)^j \frac{(m + j)!}{(j!)^2(m - j)!}.$$

But

$$\frac{(m + j)!}{(j!)^2(m - j)!} \leq (1 + 1 + 1)^{m+j} = 3^{m+j}.$$

Hence

$$|C_{ij}| \leq \sqrt{2i + 1} \, 3^{m+j}.$$

Therefore

$$\sum_{j=0}^{i} |C_{ij}| \leq (2i + 1)^{1/2} \sum_{j=0}^{i} 3^{i+j} \tag{4.32}$$

$$= \sqrt{2n + 1} \, 3^i \frac{3^{i+1} - 1}{3 - 1}$$

$$< \sqrt{2n + 1} \, \frac{3}{2} . 3^{2i}. \tag{4.33}$$

Since $C_{ij} = 0$ for $j > i$, we have

$$\sum_{p_1,..,p_d=0}^{\infty} |C_{k_1 p_1}...C_{k_d p_d}| = \left(\sum_{p_1=0}^{k_1} |C_{k_1 p_1}| \right) ... \left(\sum_{p_d=0}^{k_d} |C_{k_d p_d}| \right) \tag{4.34}$$

In view of (4.33), (4.34), we see that

$$\left(\sum_{p_1,..,p_d=0}^{\infty} |C_{k_1 p_1}...C_{k_d p_d}| \right)^2$$

$$\leq \left(\frac{3}{2}\right)^{2d} \left(\prod_{j=1}^{d}(2k_j+1)\right) 3^{4\sum_{j=1}^{d}k_i}. \tag{4.35}$$

From (4.35) one has

$$\sum_{k_1,..,k_d=0}^{n} \left(\sum_{p_1,..,p_d=0}^{\infty} |C_{k_1p_1}...C_{k_dp_d}|\right)^2$$
$$\leq \left(\frac{3}{2}\right)^{2d}(2n+1)^d \left(\sum_{j=0}^{n} 3^{4j}\right)^d.$$

Hence

$$\sum_{k_1,..,k_d=0}^{n} \left(\sum_{p_1,..,p_d=0}^{\infty} |C_{k_1p_1}...C_{k_dp_d}|\right)^2$$
$$\leq \left(\frac{3}{2}\right)^{2d}(2n+1)^d \left(\frac{3^{4(n+1)}-1}{3^4-1}\right)^d$$
$$\leq \left(\frac{3}{2}\right)^{2d}(2n+1)^d \left(\frac{81}{80}\right)^d 3^{4nd}. \tag{4.36}$$

Put

$$f(t) = \left(\frac{3}{2}\right)^d \left(\frac{81}{80}\right)^{d/2}(2t+1)^{d/2}3^{2td}$$
$$= \left(\frac{27}{8\sqrt{5}}(2t+1)^{1/2}3^{2t}\right)^d.$$

Then we have in view of (4.36)

$$\|p^{n(\epsilon)}(\mu) - p^{n(\epsilon)}(\mu^0)\| \leq \epsilon^{1/2}. \tag{4.37}$$

If we put

$$\eta(\epsilon) = \epsilon^{1/2} + \left(\sum_{\max_{1\leq i\leq d}|k_i|>n(\epsilon)} \left(\sum_{p_1,..,p_d=0}^{\infty} C_{k_1p_1}...C_{k_dp_d}\mu^0_{p_1..p_d}\right)^2\right)^{1/2}$$

then by (4.30), (4.37), it follows that

$$\|p^{n(\epsilon)}(\mu) - u_0\| \leq \eta(\epsilon).$$

As $\epsilon \to 0$, we have $f^{-1}(\epsilon^{-1/2}) \to \infty$ and $n(\epsilon) \to \infty$. By (4.15)

$$\sum_{\max_{1\leq i\leq d}|k_i|>n(\epsilon)} \left(\sum_{p_1,..,p_d=0}^{\infty} C_{k_1p_1}...C_{k_dp_d}\mu^0_{p_1..p_d}\right)^2 \to 0 \quad \text{for } \epsilon \to 0.$$

Hence $\eta(\epsilon) \to 0$ for $\epsilon \to 0$.

Finally, let $u_0 \in H^1(I)$. By Theorem 4.1 one has

$$\|p^{n(\epsilon)}(\mu^0) - u_0\| \leq \frac{1}{2(n(\epsilon) + 1)}\|u_0\|_{H^1(I)}. \tag{4.38}$$

We estimate $n(\epsilon)$. From the definition of $n(\epsilon)$, one has

$$n(\epsilon) + 1 > f^{-1}(\epsilon^{-1/2}).$$

Since f is a monotone increasing function, the foregoing inequality gives

$$f(n(\epsilon) + 1) > \epsilon^{-1/2},$$

i.e.

$$\frac{27}{8\sqrt{5}}\sqrt{2t_\epsilon + 1}\, 3^{2t_\epsilon} > \epsilon^{-1/2d}$$

with

$$t_\epsilon = n(\epsilon) + 1.$$

Since

$$t_\epsilon + 1 \leq 2^{t_\epsilon}$$

we have

$$\frac{27}{8\sqrt{5}}\sqrt{2}\, 2^{t_\epsilon/2}.3^{2t_\epsilon} > \epsilon^{-1/2}$$

which gives

$$e^{(2\ln 3 + \frac{1}{2}\ln 2)t_\epsilon} > \frac{8\sqrt{5}}{27\sqrt{2}\sqrt{\epsilon}}.$$

Hence

$$t_\epsilon \geq \left(2\ln 3 + \frac{1}{2}\ln 2\right)^{-1}\ln\left(\frac{8\sqrt{5}}{27\sqrt{2}\sqrt{\epsilon}}\right). \tag{4.39}$$

In view of (4.39), inequality (4.38) gives

$$\|p^{n(\epsilon)}(\mu^0) - u_0\| \leq \frac{1}{C(\epsilon)}\|u_0\|_{H^1(I)} \tag{4.40}$$

where

$$C(\epsilon) = 2\left(2\ln 3 + \frac{1}{2}\ln 2\right)^{-1}\ln\left(\frac{8\sqrt{5}}{27\sqrt{2}\,\sqrt[2d]{\epsilon}}\right).$$

By (4.37), (4.40), the proof of Theorem 4.2 is completed.

4.2 A moment problem from Laplace transform

We consider the problem of approximating $u_0 \in L^2(0, \infty)$ such that

$$\int_0^\infty u_0(x)e^{-kx}dx = \mu_k^0, \qquad k = 1, 2, \ldots. \tag{4.41}$$

Put $t = e^{-x}$, $w_0(t) = tu_0(-\ln t)$. It follows from (4.41) that w_0 satisfies

$$\int_0^1 w_0(t)t^k dt = \mu_{k+2}^0, \qquad k = 0, 1, 2, \ldots. \tag{4.42}$$

Note that

$$\int_0^1 |w_0(t)|^2 dt = \int_0^\infty |u_0(x)|^2 e^{-3x}dx.$$

Hence, since $u_0 \in L^2(0, \infty)$, we have $w_0 \in L^2(0, 1)$. Using Theorem 4.2 $(d = 1)$ for the system (4.42), one has the following

Theorem 4.3. *Let $u_0 \in L^2(0, \infty)$ be the solution of (4.41) corresponding to $\mu^0 = (\mu_1^0, \mu_2^0, ..) \in l^\infty$ in the right hand side of (4.41). Let*

$$f(t) = \frac{27}{8\sqrt{5}}(2t+1)^{1/2}3^{2t}$$

and for $0 < \epsilon < 1$, put

$$n(\epsilon) = [f^{-1}(\epsilon^{-1/2})].$$

Then there exists a function $\eta(\epsilon)$, $0 < \epsilon < 1$ such that $\eta(\epsilon) \to 0$ as $\epsilon \to 0$ and that for all sequences $\mu = (\mu_1, \mu_2, ..)$ satisfying

$$\|\mu - \mu^0\|_{l_\infty} \leq \epsilon$$

we have

$$\|q^{n(\epsilon)}(\mu) - u_0\|_\rho \leq \eta(\epsilon) \tag{4.43}$$

where

$$q^{n(\epsilon)}(\mu) = e^x p^{n(\epsilon)}(\tilde{\mu})(e^{-x}), \qquad \tilde{\mu} = (\mu_2, \mu_3, \ldots).$$

The norm $\| \cdot \|_\rho$ is defined by $\|h\|_\rho^2 = \int_0^\infty |h(x)|^2 e^{-3x}dx$,

and $p^{n(\epsilon)}(\tilde{\mu})$ is as in Theorem 4.2.
Moreover, if $u_0 \in H^1(0, \infty)$ then

$$\|q^{n(\epsilon)}(\mu) - u_0\|_\rho \leq \epsilon^{1/2} + \frac{3\|u_0\|_{H^1(0,\infty)}}{C(\epsilon)} \tag{4.44}$$

where $C(\epsilon)$ is as in Theorem 4.2.

Proof.

From Theorem 4.2, one gets

$$\|p^{n(\epsilon)}(\tilde{\mu}) - w_0\| \leq \eta(\epsilon), \qquad \tilde{\mu} = (\mu_2, \mu_3, ..) \tag{4.45}$$

where

$$w_0(t) = tu_0(-\ln t).$$

But

$$\|p^{n(\epsilon)}(\tilde{\mu}) - w_0\|^2 = \int_0^1 |p^{n(\epsilon)}(\tilde{\mu})(t) - w_0(t)|^2 dt$$

$$= \int_0^\infty |p^{n(\epsilon)}(\tilde{\mu})(e^{-x}) - e^{-x}u_0(x)|^2 e^{-x} dx$$

$$= \int_0^\infty |e^x p^{n(\epsilon)}(\tilde{\mu})(e^{-x}) - u(x)|^2 e^{-3x} dx$$

$$= \|q^{n(\epsilon)}(\mu) - u_0\|_\rho^2. \tag{4.46}$$

By (4.45), (4.46), inequality (4.43) holds. Now if $u_0 \in H^1(0, \infty)$ then

$$w_0' = u_0(-\ln t) - u_0'(-\ln t).$$

Hence

$$\|w_0\|_{H^1(0,1)} \leq \left(\int_0^1 |tu_0(-\ln t)|^2 dt \right)^{1/2} +$$

$$+ \left(\int_0^1 |u_0(-\ln t)|^2 dt \right)^{1/2} + \left(\int_0^1 |u_0'(-\ln t)|^2 dt \right)^{1/2}$$

$$= \left(\int_0^\infty |u_0(x)|^2 e^{-3x} dx \right)^{1/2} + \left(\int_0^\infty |u_0(x)|^2 e^{-x} dx \right)^{1/2}$$

$$+ \left(\int_0^\infty |u_0'(x)|^2 e^{-x} dx \right)^{1/2}$$

$$\leq 3\|u_0\|_{H^1(0,\infty)}. \tag{4.47}$$

From Theorem 4.2, one has

$$\|p^{n(\epsilon)}(\tilde{\mu}) - w_0\| \leq \epsilon^{1/2} + \frac{\|w_0\|_{H^1(0,1)}}{C(\epsilon)}.$$

In view of (4.46) and (4.47) the latter inequality gives

$$\|q^{n(\epsilon)}(\mu) - u_0\|_\rho \leq \epsilon^{1/2} + \frac{3\|u_0\|_{H^1(0,\infty)}}{C(\epsilon)}.$$

This completes the proof of the theorem.

4.3 Notes and remarks

1. Using Theorem 2 in [AGT], we can get an estimate which is sharper than the last one of Theorem 4.2. In fact, one has

$$\|p^{\tilde{n}(\epsilon)} - u_0\| \le \epsilon^{1/2} + \frac{\|u_0\|_{H^1(I)}}{C_1 + C_2 \ln \frac{1}{\epsilon}}$$

where

$$\tilde{n}(\epsilon) = [\tilde{f}^{-1}(\epsilon^{-1/2})], \quad \tilde{f}(t) = \frac{1}{\sqrt{(2\pi)^d}}(3 + 2\sqrt{2})^{dt+d}$$

and

$$C_1 = \frac{\ln(\sqrt{2\pi}(3 - 2\sqrt{2}))}{\ln(3 + 2\sqrt{2})} \quad and \quad C_2 = \frac{1}{2d\ln(3 + 2\sqrt{2})}.$$

Moreover, in [AGT], we also proved that the order magnitude $0(\frac{1}{\ln(1/\epsilon)})$ of the above estimate is optimal in an appropriate sense.

2. In Chapter 2 (Notes and Remarks) was presented an example showing the ill-posedness of the Hausdorff moment problem. Following is another example. Consider the sequence

$$u_n(x) = \cos 2n\pi x, \quad 0 \le x \le 1.$$

By direct computation, one has

$$\|u_n\|^2_{L^2(0,1)} = \frac{1}{2}, \quad n = 1, 2, \ldots.$$

Consider the map $A : L^2(0,1) \longrightarrow l^2$ defined by

$$(Au)_k = \int_0^1 u(x)x^k dx \quad k = 0, 1, 2, \ldots.$$

Then we have

$$|(Au_n)_k| \le \frac{1}{k+1}.$$

Since $\left(\frac{1}{k+1}\right) \in l^2$, we see that the sequence (Au_n) is bounded by a fixed element of l^2. By the dominated convergence theorem (applied to $\mathbf{N}$ with the counting measure), we have that

$$Au_n \longrightarrow 0 \text{ in } l^2.$$

Since $\|u_n\|_{L^2(0,1)} = 1/\sqrt{2} \ \forall n$, we see that A^{-1} is not continuously invertible on $R(A)$, i.e. the problem is ill-posed.

3. Although, as shown above, the Hausdorff moment problem is ill-posed, it is controllable, with A as defined in 2. above, which means that the range $R(A)$ of A is dense in l^2. Indeed, we claim that if $\mu \in l^2$ is such that

$$(\mu, Au)_{l^2} = 0 \quad \forall u \in L^2(0,1)$$

or equivalently

$$\sum_{n=0}^{\infty} \mu_k \left(\int_0^1 x^k u(x) dx \right) = 0, \qquad \forall u \in L^2(0,1), \tag{4.48}$$

then $\mu = 0$. In fact, let $v \in L^2(0,1/2)$, and let $\tilde{v}$ be the natural extension of v on $(0,1)$,

$$\begin{aligned} \tilde{v}(x) &= v(x) \quad \text{if } 0 < x < 1/2, \\ &= 0 \qquad \text{if } 1/2 < x < 1, \end{aligned} \tag{4.49}$$

then $\tilde{v} \in L^2(0,1)$ and by the condition (4.48) above

$$0 = \sum_{n=0}^{\infty} \int_0^1 x^k \mu_k \tilde{v}(x) dx = \int_0^{1/2} x^k \mu_k u(x) dx. \tag{4.50}$$

Since $|x^k| \leq 1/2^k$ on $[0,1/2]$ and since $\sup_{k \in \mathbf{N}} |\mu_k| < \infty$, the series

$$F(x) = \sum_{k=0}^{\infty} \mu_k x^k \tag{4.51}$$

converges uniformly and absolutely on $[0,1/2]$ and defines a continuous function on $[0,1/2]$. Now, interchanging the order of integration and summation in (4.51) and recalling the definition (4.51) of $F(x)$ we have

$$\int_0^{1/2} v(x) F(x) dx = 0, \qquad \forall v \in L^2(0,1/2).$$

It follows that $F(x) = 0 \ \forall x \in [0,1/2]$. As a consequence $\mu_k = 0 \ \forall k = 0,1,2,\ldots$ (i.e. $\mu = 0$). We have completed the proof that $R(A)$ is dense in l^2.

4. As shown in 3., $\overline{R(A)} = l^2$. Now, $R(A) \neq l^2$. In fact, we claim that $e_1 = (1,0,0,\ldots)$ is in $l^2 \setminus R(A)$ since there is no function in $L^2(0,1)$ such that $Au = e_1$, or equivalently, that $\int_0^1 u(x) dx = 1$ and $\int_0^1 x^k u(x) dx = 0 \ \forall k = 1,2,\ldots$.

The fact that $\overline{R(A)} = l^2$ and $R(A) \neq l^2$ can be used to prove that A^{-1} is unbounded on $R(A)$ to $L^2(0,1)$, i.e., that the Hausdorff moment problem is ill-posed.

5. This section is devoted to a regularization of the Hausdorff moment problem by the Backus-Gilbert method (BG method for short).

Consider the Hausdorff moment problem on $(0,1)$.

Find u in $L^p(0,1)$, $1 < p < \infty$, satisfying

$$\int_0^1 x^k u(x) dx = \mu_k, \qquad k = 0,1,2,\ldots.$$

Following the BG method, we approximate the solution $u \in L^p(0,1)$ of the moment problem by finite combinations

$$u^n(\mu) = \sum_{j=1}^{n} \mu_j v_j^n$$

where v_j^n, $j = 1, \ldots, n$, are the BG basis functions. The construction of these basis functions, we recall, is done according to the following procedure:

For $x \in \mathbf{R}$, the value of v_j^n at x is defined as a solution of the following minimization problem

$$\min_{v \in L_n} S_x^n(v)$$

where L_n is the hyperplane

$$L_n = \left\{ v = (v_1, v_2, \ldots, v_n) \in \mathbf{R}^n : \sum_{j=1}^n \frac{v_j}{j} = 1 \right\}$$

and

$$S_x^n(v) = \int_0^1 |x - y|^\beta \left| \sum_{j=1}^n v_j(y) y^j \right|^q dy.$$

Here $n \in \mathbf{N}$ and $\beta > 0$ is fixed.

Now L_n is closed and convex and by the results of Chapter 3, S_x^n is strictly convex, continuous and coercive. Hence the above minimization problem has a unique solution $v^n(x) \in L_n$ such that

$$S_x^n(v^n(x)) = \min_{w \in L_n} S_x^n(w).$$

Defining $v^n(x)$ for each $x \in \mathbf{R}$ by this procedure, we obtain the basis functions $v^n = (v_1^n, \ldots, v_n^n)$ on $\mathbf{R}$. One can show that the v_n's are continuous on $\mathbf{R}$. For $\sigma \in [0, 1)$ and $q \in [0, \infty)$, let $(W^{\sigma,q}(0,1), \| \cdot \|_{\sigma,q})$ be the fractional Sobolev space defined as in Chapter 3. Then we have

Proposition 4.4. *Let u_0 be the solution of the Hausdorff moment problem above, associated with a sequence $\mu^0 = (\mu_0^0, \mu_1^0, \ldots)$.*

Let

$$\delta_0 = \left(\min_{0 \leq x \leq 1} \int_0^1 |x - y|^\beta dy \right)^{2/q}$$

$$C(n) = \min\{S_x^n(v) |\ x \in [0,1],\ |v| = \sum_{j=0}^{n-1} |v_j| = 1\}$$

and choose, for $\delta \in (0, \delta_0)$,

$$n(\delta) = [f^{-1}(\delta^{-1/2})] \in \mathbf{N}$$

where f is chosen to be continuous and strictly increasing on $[0, \infty)$ to $[(C(0))^{-1/q}, \infty)$ such that

$$f(n) \geq (C(n))^{-1/q} \qquad \forall n \in \mathbf{N}.$$

a) Suppose $u \in W^{\sigma,\infty}(0,1)$, $\sigma \in (0,1]$ and

$$q - 1 \leq \beta < (\sigma + 1)q - 1, \quad \|\mu - \mu^0\|_\infty \leq \delta.$$

Then the following estimate of the error between the exact solution and the finite dimensional approximation $u^{n(\delta)}$ holds:

$$\|u^{n(\delta)}(\mu) - u_0\|_{L^\infty} \leq$$
$$C\left(\delta^{1/2} + \|u\|_{\sigma,\infty} \begin{cases} [f^{-1}(\delta^{-1/2})]^{-s/p} & \text{if } \beta > q - 1 \\ \ln[f^{-1}(\delta^{-1/2})]^{-1/p} & \text{if } \beta = q - 1 \end{cases}\right)$$

where $0 < s < \frac{\beta}{q-1} - 1$ and C depends only on q, β, σ and s.

b) Let $u_0 \in W^{\cdot\sigma, max(\gamma,p)}(0,1)$ with $\sigma \in [0,1)$, $\gamma \in [1,\infty)$ and let

$$q - 1 \leq \beta < (\sigma + 1)q - 1 \text{ if } \gamma > \delta$$

or

$$q - 1 \leq \beta \leq (\sigma + 1)q - 1 \text{ if } \gamma \leq \delta.$$

Suppose

$$\|\mu - \mu^0\|_{L^\infty} \leq \delta.$$

Then we have the following error estimate in the L^γ-norm

$$\|u^{n(\delta)}(\mu) - u_0\|_{L^\gamma} \leq$$
$$C\left(\delta^{1/2} + \|u\|_{\sigma, max(\gamma,p)} \begin{cases} [f^{-1}(\delta^{-1/2})]^{-s/p} & \text{if } \beta > q - 1 \\ \ln[f^{-1}(\delta^{-1/2})]^{-1/p} & \text{if } \beta = q - 1 \end{cases}\right).$$

Here s is as in a) and depends only on q, β, γ, δ and s.

The above proposition is a consequence of Theorem 3.7 since $(0,1)$ satisfies the uniform cone condition. In the above proposition, there may be no solution to (MP) associated with μ. In the case a solution corresponding to μ does exist, we can obtain a sharper error estimate. We do not pursue the matter any further but instead, leave it to the reader to derive approximate estimates, using the general results of Chapter 3.

5 Analytic functions: reconstruction and Sinc approximations

In this chapter, we address two problems in function theory:

- The problem of reconstructing an analytic function in the Hardy space $H^2(U)$ of the unit disc from a sequence of moments.

- A cardinal series representation theorem in the two dimensional case.

Both problems can be viewed as moment problems. The results of this chapter, beside their intrinsic interest, will play an important role in our regularization of the inverse problems in Potential Theory and in Heat Conduction to be considered in the chapters to follow.

As is known, the problem of reconstructing an analytic function from its values at a sequence of points of its domain is an ill-posed problem. Two approaches will be followed: the first one is by polynomial approximation, and the second one is through optimal recovery.

In the case of analytic functions of one complex variable, it follows from the Paley-Wiener theorem that a function in $L^2(\mathbf{R})$ admits a cardinal series representation iff its Fourier transform has bounded support. For functions in $L^2(\mathbf{R}^2)$, whose Fourier transforms have bounded supports, we shall establish a theorem of representation in terms of one-dimensional Sinc functions. This Sinc representation theorem will be used in our approximation of functions in $L^2(\mathbf{R}^2)$.

All function spaces in this chapter are complex function spaces. For notational simplicity, we use the same notations as in previous chapters to denote function spaces. Similarly, all sequences in this chapter are complex sequences.

The remainder of this chapter consists of three sections. Section 5.1 is concerned with a reconstruction of functions in $H^2(U)$ using approximating polynomials. Section 5.2 deals with the same problem, using the approach of optimal recovery. The final Section 5.3 is devoted to two-dimensional Sinc theory and a moment problem on $\mathbf{R}^2$.

5.1 Reconstruction of functions in $H^2(U)$: approximation by polynomials

Let U be the open unit disc of $\mathbf{C}$. We consider the problem of reconstructing a function u in $H(U)$ (the space of functions analytic on U) satisfying the following moment problem

$$u(z_n) = \mu_n, \qquad n = 1, 2, ..., \tag{5.1}$$

where (z_n) is a sequence in U such that $z_n \neq z_m$ for $n \neq m$, $z_n \to 0$ as $n \to \infty$ and $(\mu_n) \in l^\infty$.

For definiteness, we take

$$z_n = \frac{1}{n+1}, \qquad n = 1, 2, \ldots .$$

Let $\mu^0 = (\mu_n^0) \in l^\infty$ be a sequence such that the following problem has a (unique) solution u_0 in $H(U)$

$$u_0(z_n) = \mu_n^0. \tag{5.2}$$

Let $\mu = (\mu_n) \in l^\infty$ satisfy

$$\|\mu - \mu^0\|_\infty = \sup_n |\mu_n - \mu_n^0| < \epsilon.$$

We shall construct a polynomial that, in a sense to be specified, approximates the exact solution u_0. For $m \in \mathbf{N}$, we consider the following system of linear equations

$$\sum_{k=0}^{m-1} a_{mk} z_n^k = \mu_n, \qquad n = 1, 2, .., m. \tag{5.3}$$

Put

$$P_m(z) = \sum_{0 \le k \le m/2} a_{mk} z^k. \tag{5.4}$$

We shall give an estimate of the error between P_n and the exact solution u_0 corresponding to the exact data $\mu^0 \in l^\infty$.

Before stating our main result, we set some notations. The Hardy space $H^2(U)$ is the class of all analytic functions F on U having the form

$$F(z) = \sum_{k=0}^\infty \alpha_k z^k$$

with

$$\sum_{k=0}^\infty |\alpha_k|^2 < \infty.$$

Writing H^2 for $H^2(U)$, we define the norm

$$\|F\|_{H^2} = \left(\sum_{k=0}^\infty |\alpha_k|^2 \right)^{1/2}$$

(see, e.g. [Ho]).

We now state the main result of this section.

Theorem 5.1. *Let $u_0 \in H^2(U)$ be the solution of (5.2) corresponding to $\mu^0 = (\mu_n^0)$. Let*

$$f(\theta) = (\theta + 2) \left(\frac{32e(\theta+1)^2}{\theta} \right)^\theta, \qquad \theta > 0,$$

and for $0 < \epsilon < 1$, put

$$m(\epsilon) = [f^{-1}(\epsilon^{-1})]$$

where $[x]$ is the largest integer $\leq x$.

Then there exists a function $\eta(\epsilon)$, $0 < \epsilon < 1$ such that $\eta(\epsilon) \to 0$ as $\epsilon \to 0$ and that, for all sequences $\mu = (\mu_n) \in l^\infty$ satisfying

$$\|\mu - \mu^0\|_\infty = \sup_n |\mu_n - \mu_n^0| < \epsilon,$$

we have

$$\|u_0 - P_{m(\epsilon)}\|_{H^2}^2 \leq \eta(\epsilon)$$

where $P_{m(\epsilon)}$ is defined by (5.4) for $m = m(\epsilon)$.

If, in addition $u_0' \in H^2(U)$, then for $\epsilon \to 0$,

$$\|u_0 - P_{m(\epsilon)}\|_{H^2}^2 \leq \delta(\epsilon)$$

where

$$\delta(\epsilon) = \epsilon + 16\|u_0\|_{H^2}^2 (m(\epsilon) + 2) \left(\frac{32e}{m(\epsilon)}\right)^{m(\epsilon)} + \frac{4\|u_0'\|_{H^2}^2}{m^2(\epsilon)}.$$

Proof. Since $u_0 \in H^2(U)$, u_0 is represented by a series

$$u_0(z) = \sum_{k=0}^\infty a_k z^k \tag{5.5}$$

where

$$\sum_{k=0}^\infty |a_k|^2 < \infty. \tag{5.6}$$

On the other hand,

$$u_0(z_n) = \mu_n^0, \qquad n = 1, 2, \ldots .$$

Hence, (5.5) gives

$$\sum_{k=0}^{m-1} a_k z_n^k = \mu_n^0 - \sum_{k=m}^\infty a_k z_n^k. \tag{5.7}$$

Subtracting (5.7) from (5.3) gives

$$\sum_{k=0}^{m-1} c_{mk} z_n^k = \varphi_{mn}, \qquad n = 1, 2, .., m, \tag{5.8}$$

where

$$c_{mk} = a_{mk} - a_k, \qquad k = 0, 1, ..m-1,$$

$$\varphi_{mn} = \epsilon_n + \sum_{k=m}^\infty a_k z_n^k,$$

$$\epsilon_n = \mu_n - \mu_n^0, \qquad n = 1, 2, .., m.$$

We wish to estimate $|c_{mk}|$. For this purpose, we shall give an explicit form of (c_{mk}). Put

$$Q_m(z) = \sum_{k=0}^{m-1} c_{mk} z^k. \tag{5.9}$$

Eq. (5.8) gives

$$Q_m(z_n) = \varphi_{mn}, \qquad n = 1, 2, .., m.$$

Hence, using Lagrange's interpolation formula (cf. [MSM]), we get:

$$Q_m(z) = \sum_{n=1}^{m} \varphi_{mn} \frac{(z - z_1)..(z - z_{n-1})(z - z_{n+1})..(z - z_m)}{(z_n - z_1)..(z_n - z_{n-1})(z_n - z_{n+1})..(z_n - z_m)}. \tag{5.10}$$

We shall rewrite (5.10) in a more convenient form. Let us first set some notations. For $j = 1, 2, .., m$, put

$$\hat{z}_j = (z_1, .., z_{j-1}, z_{j+1}, .., z_m) \in \mathbf{R}^{m-1}.$$

Letting $t = (t_1, .., t_{m-1}) \in \mathbf{R}^{m-1}$, we define the elementary symmetric functions σ_j ($j = 1, 2, .., m - 1$) (see [MSM]) of t by

$$\sigma_0(t) = 1,$$
$$\sigma_1(t) = t_1 + t_2 + .. + t_{m-1},$$
$$\sigma_2(t) = \sum_{1 \le j_1 < j_2 \le m-1} t_{j_1} t_{j_2},$$
$$\sigma_3(t) = \sum_{1 \le j_1 < j_2 < j_3 \le m-1} t_{j_1} t_{j_2} t_{j_3},$$
$$\vdots \quad \vdots$$
$$\sigma_{m-1}(t) = t_1 t_2 .. t_{m-1}.$$

We next set

$$s_{mn} = (z_n - z_1)..(z_n - z_{n-1})(z_n - z_{n+1})..(z_n - z_m).$$

Using these notations, we can rewrite (5.10) as

$$Q_m(z) = \sum_{n=1}^{m} \varphi_{mn} s_{mn}^{-1} \sum_{k=0}^{m-1} z^k (-1)^{m-k-1} \sigma_{m-k-1}(\hat{z}_n)$$
$$= \sum_{k=0}^{m-1} \left(\sum_{n=1}^{m} \varphi_{mn} s_{mn}^{-1} (-1)^{m-k-1} \sigma_{m-k-1}(\hat{z}_n) \right) z^k. \tag{5.11}$$

Combining (5.9) with (5.11) gives

$$c_{mk} = \sum_{n=1}^{m} \varphi_{mn} s_{mn}^{-1} (-1)^{m-k-1} \sigma_{m-k-1}(\hat{z}_n). \tag{5.12}$$

Now, we have

$$|\varphi_{mn}| \leq |\epsilon_n| + \sum_{k=m}^{\infty} |a_k||z_n|^k.$$

Noting that $|a_k| \leq \|u_0\|_{H^2}$, we get after some computations:

$$|\varphi_{mn}| \leq \epsilon + \frac{\|u_0\|_{H^2}|z_n|^m}{1 - |z_n|}.$$

Since $z_n = 1/(n+1) \leq 1/2$, $n = 1, 2, ..$, the latter inequality gives

$$|\varphi_{mn}| \leq \epsilon + \frac{2\|u_0\|_{H^2}}{(n+1)^m}. \tag{5.13}$$

We propose to estimate $|s_{mn}^{-1}|$. We have

$$\begin{aligned}
s_{mn} &= (z_n - z_1)..(z_n - z_{n-1})(z_n - z_{n+1})..(z_n - z_m) \\
&= \left(\frac{1}{n+1} - \frac{1}{2}\right)\left(\frac{1}{n+1} - \frac{1}{3}\right)\cdots\left(\frac{1}{n+1} - \frac{1}{n}\right)\left(\frac{1}{n+1} - \frac{1}{n+2}\right)\cdots \\
&\quad \cdots\left(\frac{1}{n+1} - \frac{1}{m+1}\right) \\
&= \frac{(-1)^{n-1}(n-1)!(m-n)!}{(n+1)^{m-2}(m+1)!}.
\end{aligned}$$

Hence

$$|s_{mn}|^{-1} = \frac{(n+1)^{m-2}(m+1)!}{(n-1)!(m-n)!}. \tag{5.14}$$

To estimate $|\sigma_{m-k-1}(\hat{z}_k)|$ we note that if

$$t_1 > t_2 > .. > t_{m-1} > 0$$

then

$$\begin{aligned}
|\sigma_{m-k-1}(t)| &= \left| \sum_{1 \leq j_1 < .. < j_{m-k-1} \leq m-1} t_{j_1} t_{j_2} .. t_{j_{m-k-1}} \right| \\
&\leq t_1 t_2 .. t_{m-k-1} \left| \sum_{1 \leq j_1 < .. < j_{m-k-1} \leq m-1} 1 \right|, \tag{5.15}
\end{aligned}$$

$t = (t_1, .., t_m)$. Note that the quantity

$$\sum_{1 \leq j_1 < .. < j_{m-k-1} \leq m-1} 1$$

is the number of $(m-k-1)$-element subsets of the set $\{1, 2, .., m-1\}$. Hence

$$\sum_{1 \leq j_1 < .. < j_{m-k-1} \leq m-1} 1 = \binom{m-1}{m-k-1}. \tag{5.16}$$

We get in view of (5.15), (5.16)

$$|\sigma_{m-k-1}(t)| \le t_1 t_2 .. t_{m-k-1} \binom{m-1}{m-k-1}.$$

For $t = \hat{z}_n$, the latter inequality gives

$$|\sigma_{m-k-1}(\hat{z}_n)| \le \frac{1}{(m-k)!} \binom{m-1}{m-k-1}. \tag{5.17}$$

In view of (5.13), (5.14), (5.17), the relation (5.12) implies

$$
\begin{aligned}
|c_{mk}| &\le \sum_{n=1}^{m} \left(\epsilon + \frac{2\|u_0\|_{H^2}}{(n+1)^m} \right) \binom{m-1}{m-k-1} \times \\
&\qquad \times \frac{1}{(m-k)!} \cdot \frac{(n+1)^{m-2}(m+1)!}{(n-1)!(m-n)!} \\
&\le \frac{\epsilon \binom{m-1}{m-k-1}}{(m-k)!} \sum_{n=1}^{m} \frac{(n+1)^{m-2}(m+1)!}{(n-1)!(m-n)!} + \\
&\quad + \frac{2\|u_0\|_{H^2} \binom{m-1}{m-k-1}}{(m-k)!} \sum_{n=1}^{m} \frac{(m+1)!}{(n+1)^2(n-1)!(m-n)!}. \tag{5.18}
\end{aligned}
$$

But, we have

$$\frac{(m+1)!}{(n+1)^2(n-1)!(m-n)!} \le \frac{(m+1)!}{(n+1)!(m-n)!} = \binom{m+1}{n+1}. \tag{5.19}$$

On the other hand,

$$\sum_{k=0}^{p} \binom{p}{k} = (1+1)^p = 2^p, \qquad p = 1, 2, \ldots. \tag{5.20}$$

From (5.18), (5.19), (5.20) we get after some computations

$$
\begin{aligned}
|c_{mk}| &\le \frac{\epsilon 2^{m-1}(m+1)^m}{(m-k)!} \sum_{n=1}^{m} \binom{m+1}{n+1} \\
&\quad + \frac{2^m \|u_0\|_{H^2}}{(m-k)!} \sum_{n=1}^{m} \binom{m+1}{n+1} \\
&\le \frac{\epsilon 4^m (m+1)^m}{(m-k)!} + \frac{4^{m+1}\|u_0\|_{H^2}}{(m-k)!}, \tag{5.21}
\end{aligned}
$$

$$k = 0, 1, .., m-1.$$

Stirling's formula (cf. [Tay]) gives

$$(m-k)! = \left(\frac{m-k}{e} \right)^{m-k} \sqrt{2\pi(m-k)} e^{r_k}$$

where
$$\frac{1}{12(m-k)} < r_k < \frac{1}{12(m-k-1)}.$$

Hence
$$(m-k)! \geq \left(\frac{m-k}{e}\right)^{m-k}.$$

For $0 \leq k \leq m/2$, one has
$$(m-k)! \geq \left(\frac{m-k}{e}\right)^{m-k} \geq \left(\frac{m}{2e}\right)^{m/2}, \quad m = 1, 2, \ldots . \tag{5.22}$$

Substituting (5.22) into (5.22), one gets, for $0 \leq k \leq m/2$,
$$|c_{mk}| \leq \epsilon \left(\frac{4(m+1)\sqrt{2e}}{\sqrt{m}}\right)^m + 4\|u_0\|_{H^2} \left(\frac{4\sqrt{2e}}{\sqrt{m}}\right)^m. \tag{5.23}$$

Now, we have
$$P_m(z) - u_0(z) = \sum_{0 \leq k \leq m/2} c_{mk} z^k + \sum_{k > m/2} a_k z^k$$

where
$$P_m(z) = \sum_{0 \leq k \leq m/2} a_{mk} z^k.$$

Hence
$$\|P_m - u_0\|_{H^2}^2 = \sum_{0 \leq k \leq m/2} |c_{mk}|^2 + \sum_{k > m/2} |a_k|^2.$$

Using (5.23) and the inequality
$$(a+b)^2 \leq 2(a^2 + b^2), \quad \forall a, b \in \mathbf{R},$$

we get
$$\|P_m - u_0\|_{H^2}^2 \leq 2\left(\frac{m}{2}+1\right)\epsilon^2 \left(\frac{32e(m+1)^2}{m}\right)^m + 16(m+2)\|u_0\|_{H^2}^2 \left(\frac{32e}{m}\right)^m$$

$$+ \sum_{k \geq m/2} |a_k|^2. \tag{5.24}$$

Now choosing
$$m(\epsilon) = [f^{-1}(\epsilon^{-1})]$$

with
$$f(\theta) = (\theta + 2)\left(\frac{32e(\theta+1)^2}{\theta}\right)^\theta$$

we have

$$(m(\epsilon) + 2) \left(\frac{32e(m(\epsilon) + 1)^2}{m(\epsilon)} \right)^{m(\epsilon)} \leq \epsilon^{-1}.$$

Hence, (5.24) implies

$$\|P_{m(\epsilon)} - u_0\|_{H^2}^2 \leq \epsilon + 16\|u_0\|_{H^2}^2 (m(\epsilon) + 2) \left(\frac{32e}{m(\epsilon)} \right)^{m(\epsilon)} +$$

$$+ \sum_{k \geq m(\epsilon)/2} |a_k|^2 \equiv \eta(\epsilon). \tag{5.25}$$

Now, if $u_0' \in H^2(U)$ then

$$\sum_{k=0}^{\infty} k^2 |a_k|^2 = \|u_0'\|_{H^2}^2 < \infty.$$

Hence

$$\sum_{k \geq m(\epsilon)/2} |a_k|^2 \leq \frac{4}{m^2(\epsilon)} \sum_{k \geq m(\epsilon)/2} k^2 |a_k|^2$$

$$\leq \frac{4\|u_0'\|_{H^2}^2}{m^2(\epsilon)}. \tag{5.26}$$

From (5.22), (5.23), we get

$$\|P_{m(\epsilon)} - u_0\|_{H^2}^2 \leq \epsilon + 16\|u_0\|_{H^2}^2 (m(\epsilon) + 2) \left(\frac{32e}{m(\epsilon)} \right)^{m(\epsilon)} +$$

$$+ \frac{4\|u_0'\|_{H^2}^2}{m^2(\epsilon)}.$$

This completes the proof of Theorem 5.1.

5.2 Reconstruction of an analytic function: a problem of optimal recovery

In the preceding section, we considered the problem of recovering an analytic function in $H^2(U)$ from its values on the sequence of points $z_n = (n+1)^{-1}$, $n = 1, 2, \ldots$. If (z_n) is an infinite sequence of points in U, then a regularization can be performed using any of the methods presented in previous chapters. However, to obtain error estimates is a more subtle problem. In this chapter, following an alternate approach, we shall find functions $b_{m1}(z), \ldots, b_{mm}(z)$ $(m \in \mathbf{N})$ such that

$$b_{m1}(z)\mu_1 + \ldots + b_{mm}(z)\mu_m \to u \quad \text{as } m \to \infty.$$

To this end, for each $z \in U$, we put

$$\varphi_z(\mathbf{x}) = \sum_{k=0}^{\infty} \alpha_k z^k, \quad \mathbf{x} = (\alpha_k)_{k \geq 0} \in l^2.$$

Then $\varphi_z : l^2 \to \mathbf{C}$ is a continuous linear mapping. Moreover, if the function u has the expansion

$$u(z) = \sum_{k=0}^{\infty} a_k z^k,$$

then, putting $\mathbf{x}_0 = (a_k)_{k \geq 0}$, one has

$$\varphi_z(\mathbf{x}_0) = u(z), \qquad \forall z \in U.$$

Hence

$$\left| \varphi_z(\mathbf{x}_0) - \sum_{j=1}^{m} b_{mj}(z) \sum_{k=0}^{\infty} a_k z_j^k \right| = \left| u(z) - \sum_{j=1}^{m} b_{mj}(z) \sum_{k=0}^{\infty} a_k z_j^k \right|.$$

Therefore, our problem here is to find $b_{m1}(z), .., b_{mm}(z)$ such that the sequence

$$\left| \varphi_z(\mathbf{x}) - \sum_{j=1}^{m} b_{mj}(z) \sum_{k=0}^{\infty} \alpha_k z_j^k \right|, \qquad \mathbf{x} = (\alpha_k)_{k \geq 0},$$

tends to 0 as fast as possible as $m \to \infty$ for $\mathbf{x}$ in an appropriately chosen bounded subset of l^2. In the latter form, our problem can be seen as one of optimal recovery.

The method is based on the following

Proposition 5.2. *Let H be a complex Hilbert space, and let*

$$\varphi : H \to \mathbf{C}, \quad \Phi : H \to \mathbf{C}^m$$

be linear continuous mappings, $\Phi = (F_1, .., F_m)$. Then
a) there is $b_m = (b_{m1}, .., b_{mm}) \in \mathbf{C}^m$ such that

$$\varphi(\mathbf{x}) - \sum_{j=1}^{m} b_{mj} F_j(\mathbf{x}) = 0, \qquad \forall \mathbf{x} \in (\ker \Phi)^{\perp}, \tag{5.27}$$

where

$$\ker f = \{\mathbf{x} \in H : f(\mathbf{x}) = 0\}, \qquad f \in H^*,$$
$$W^{\perp} = \{\mathbf{x} \in H : (\mathbf{x}, \mathbf{y}) = 0 \ \forall \mathbf{y} \in W\}, \qquad W \subset H,$$

and $(.,.)$, H^* are the inner product and the dual of H respectively.
b) For any $b_m = (b_{m1}, .., b_{mm}) \in \mathbf{C}^m$ satisfying (5.27) one has

$$\left\| \varphi - \sum_{j=1}^{m} b_{mj} F_j \right\|_{H^*} = \sup_{x \in B_1(0) \cap \ker \Phi} |\varphi(\mathbf{x})|$$

where $B_r(0)$ $(r > 0)$ is the open ball of radius r in H centered at $0 \in H$.

Margaril-Il'yaev and Le [ML] (see also [Le1], [Le2]) proved a theorem for optimal recovery of functionals in general linear spaces. In the case that the linear space is a Hilbert space, the above proposition gives a stronger result and moreover it has a constructive character. In fact, since the (complex) dimensions dim $(\ker F_j)^\perp = 1$, $j = 1, 2, .., m$, and since

$$(\ker \Phi)^\perp = (\ker F_1)^\perp + ... + (\ker F_m)^\perp$$

we have dim $(\ker \Phi)^\perp \leq m$. Hence, in Part a) of Proposition 5.2, we can find the quantities $b_m = (b_{m1}, .., b_{mm}) \in \mathbf{C}^m$ by solving a finite system of linear algebraic equations.

Proof of Proposition 5.2.

a) By the representation theorem for linear functionals in Hilbert spaces, we can find elements ϕ, $\phi_1, ..., \phi_m$ in H such that

$$\varphi(\mathbf{x}) = (\mathbf{x}, \phi), \quad F_j(\mathbf{x}) = (\mathbf{x}, \phi_j) \tag{5.28}$$

where $(.,.)$ is the inner product in H.

From (5.28), one has $\mathbf{x} \in \ker \Phi$ iff

$$(\mathbf{x}, \phi_j) = 0, \quad j = 1, 2, .., m.$$

Hence

$$(\ker \Phi)^\perp = < \phi_1, .., \phi_m > \tag{5.29}$$

where $< \phi_1, .., \phi_m >$ is the linear space generated by $\{\phi_1, .., \phi_m\}$.

Using (5.28), we can rewrite (5.27) in the form

$$\left(\mathbf{x}, \phi - \sum_{j=1}^{m} \bar{b}_{mj} \phi_j \right) = 0, \quad \forall \mathbf{x} \in (\ker \Phi)^\perp, \tag{5.30}$$

Noting that

$$H = \ker \Phi \oplus (\ker \Phi)^\perp \tag{5.31}$$

one has

$$\phi = \phi' + \phi\prime\prime \quad \text{where } \phi' \in \ker \Phi, \ \phi\prime\prime \in (\ker \Phi)^\perp.$$

Substituting the latter relations into (5.30), one gets

$$\left(\mathbf{x}, \phi\prime\prime - \sum_{j=1}^{m} \bar{b}_{mj} \phi_j \right) = 0, \quad \forall \mathbf{x} \in (\ker \Phi)^\perp. \tag{5.32}$$

Hence (5.27) holds iff there is $b_m = (b_{m1}, .., b_{mm})$ such that

$$\phi\prime\prime = \sum_{j=1}^{m} \bar{b}_{mj} \phi_j. \tag{5.33}$$

Since $\phi_{II} \in (\ker \Phi)^{\perp}$, this gives in view of (5.29)

$$\phi_{II} \in < \phi_1, .., \phi_m > .$$

Therefore, there are $\beta_1, .., \beta_m$ such that

$$\phi_{II} = \sum_{j=1}^{m} \beta_{mj} \phi_j.$$

Putting

$$b_m = (\bar{\beta}_{m1}, .., \bar{\beta}_{mm})$$

one then has (5.33). This completes the proof of Part a).

b) For any $\mathbf{y} \in B_1(0)$, we can, in view of (5.31), write

$$\mathbf{y} = \mathbf{y}_1 + \mathbf{y}_2 \qquad \mathbf{y}_1 \in \ker \Phi, \; \mathbf{y}_2 \in (\ker \Phi)^{\perp}.$$

Since $\mathbf{y}_1$, $\mathbf{y}_2$ are mutually orthogonal, one has

$$\|\mathbf{y}\|_H^2 = \|\mathbf{y}_1\|_H^2 + \|\mathbf{y}_2\|_H^2.$$

Hence

$$\|\mathbf{y}_1\|_H^2 \leq \|\mathbf{y}\|_H^2 < 1$$

which gives

$$\mathbf{y}_1 \in \ker \Phi \cap B_1(0). \tag{5.34}$$

Now, one has

$$\varphi(\mathbf{y}) - \sum_{j=1}^{m} b_{mj} F_j(\mathbf{y}) =$$

$$= \varphi(\mathbf{y}_1) + \left(\varphi(\mathbf{y}_2) - \sum_{j=1}^{m} b_{mj} F_j(\mathbf{y}_2) \right). \tag{5.35}$$

(Here we use the fact $\mathbf{y}_1 \in \ker \Phi$, i.e., $F_j(\mathbf{y}_1) = 0, \; j = 1, 2, .., m$).

On the other hand, $b_m = (b_{m1}, .., b_{mm})$ satisfies (5.27), and, hence

$$\varphi(\mathbf{y}_2) - \sum_{j=1}^{m} b_{mj} F_j(\mathbf{y}_2) = 0. \tag{5.36}$$

From (5.35), (5.36), we get in view of (5.34)

$$|\varphi(\mathbf{y}) - \sum_{j=1}^{m} b_{mj} F_j(\mathbf{y})| = |\varphi(\mathbf{y}_1)|$$

$$\leq \sup_{\mathbf{x} \in \ker \Phi \cap B_1(0)} |\varphi(\mathbf{x})|.$$

Since $\mathbf{y}$ is any element in $B_1(0)$, the latter inequality implies

$$\|\varphi - \sum_{j=1}^{m} b_{mj} F_j\|_{H^*} \leq \sup_{\mathbf{x} \in \ker \Phi \cap B_1(0)} |\varphi(\mathbf{x})|. \tag{5.37}$$

Conversely, for $\mathbf{x} \in \ker \Phi \cap B_1(0)$, one has

$$|\varphi(\mathbf{x})| = |\varphi(\mathbf{x}) - \sum_{j=1}^{m} b_{mj} F_j(\mathbf{x})|$$

$$\leq \|\varphi - \sum_{j=1}^{m} b_{mj} F_j\|_{H^*}.$$

It follows that

$$\sup_{\mathbf{x} \in \ker \Phi \cap B_1(0)} |\varphi(\mathbf{x})| \leq \|\varphi - \sum_{j=1}^{m} b_{mj} F_j\|_{H^*}. \tag{5.38}$$

Combining (5.37), (5.38) completes the proof of Proposition 5.2.

We shall consider the case that the Hilbert space H is the set of all complex sequences $(\alpha_k)_{k \geq 0}$ such that

$$\|(\alpha_k)\|_H \equiv \left(|\alpha_0|^2 + \sum_{k=1}^{\infty} k^2 |\alpha_k|^2\right)^{1/2} < \infty.$$

On H, we consider the linear forms

$$F_n(\mathbf{x}) = \sum_{k=0}^{\infty} \alpha_k z_n^k, \quad \mathbf{x} = (\alpha_k)_{k \geq 0}, \ n = 1, 2, ...,$$

$$\Phi = (F_1, .., F_m). \tag{5.39}$$

We recall that

$$\varphi_z(\mathbf{x}) = \sum_{k=0}^{\infty} \alpha_k z^k, \quad \mathbf{x} = (\alpha_k)_{k \geq 0} \in H,$$

where (z_n) is a given sequence in U and z is a point of U.

In our newly introduced notations, the reconstruction problem (5.1) is equivalent to one of solving an infinite system of linear algebraic equations in infinitely many unknowns $\alpha_0, \alpha_1, ..$, that is,

$$F_n(\mathbf{x}) = \mu_n, \quad n = 1, 2, ... \tag{5.40}$$

Our aim is to find an estimate of the asymptotic behavior for $m \to \infty$ of the quantities

$$\sup_{\mathbf{x} \in B_{C_0}(0)} \left| \varphi_z(\mathbf{x}) - \sum_{j=1}^{m} b_{mj}(z) F_j(\mathbf{x}) \right| = C_0 \left\| \varphi_z - \sum_{j=1}^{m} b_{mj}(z) \sum_{k=0}^{\infty} \alpha_k z_j^k \right\|_{H}.$$

where $b(z) = (b_{m1}(z), .., b_{mm}(z))$ is a solution of

$$\varphi_z(\mathbf{x}) - \sum_{j=1}^{m} b_{mj}(z) F_j(\mathbf{x}) = \varphi_z(\mathbf{x}) - \sum_{j=1}^{m} b_{mj}(z) \sum_{k=0}^{\infty} \alpha_k z_j^k$$

$$= 0, \quad \forall \mathbf{x} = (\alpha_k)_{k \geq 0} \in (\ker \Phi)^{\perp}. \tag{5.41}$$

In other words, we want to estimate the degree of accuracy of finite dimensional approximations to equations (5.40).

In fact, we have

Theorem 5.3. *Let $z \in U$, let $\alpha > 0$ and $(z_j) \subset U$ satisfy*

$$z_k \neq z_j \quad \forall k \neq j, \quad |z_j| \leq \alpha \leq 1/16, \quad k, j = 1, 2, \dots.$$

Then for $C_0 > 0$, we have

$$\sup_{\mathbf{x} \in B_{C_0}(0)} \left| \varphi_z(\mathbf{x}) - \sum_{j=1}^{m} b_{mj}(z) \sum_{k=0}^{\infty} \alpha_k z_j^k \right| \leq$$

$$\leq \frac{4C_0}{1 - |z|} \max \left\{ \frac{1}{7.2^{m/2}}, \frac{2}{m} \right\}, \quad \forall z \in U.$$

provided $b_m(z) = (b_{m1}(z), .., b_{mm}(z))$ satisfies (5.41) and $\mathbf{x} = (\alpha_k)_{k \geq 0} \in B_{C_0}(0)$. If $u \in H^2(U)$, $u' \in H^2(U)$ and

$$u(z_j) = \mu_j, \quad j = 1, 2, \dots,$$

then the following holds

$$\left| u(z) - \sum_{j=1}^{m} b_{mj}(z) \mu_j \right| \leq \frac{4C_0'}{1 - |z|} \max \left\{ \frac{1}{7.2^{m/2}}, \frac{2}{m} \right\}, \quad \forall z \in U,$$

where

$$C_0' = \|u\|_{H^2(U)} + \|u'\|_{H^2(U)}.$$

Moreover, if K is a compact subset of U then

$$\sup_{z \in K} \left| u(z) - \sum_{j=1}^{m} b_{mj}(z) \mu_j \right| \leq \frac{4C_0'}{dist(K, \partial U)} \max \left\{ \frac{1}{7.2^{m/2}}, \frac{2}{m} \right\}.$$

Proof. For the proof, we shall apply Proposition 5.2. It is therefore sufficient to give an estimate of

$$\sup_{\mathbf{x} \in B_1(0) \cap \ker \Phi} |\varphi_z(\mathbf{x})|$$

where $\Phi = (F_1, .., F_m)$ as in (5.39).

From the definition, one has

$$\varphi_z(\mathbf{x}) = \sum_{k=0}^{\infty} \alpha_k z^k, \qquad \mathbf{x} = (\alpha_k)_{k \geq 0}.$$

Hence

$$|\varphi_z(\mathbf{x})| \leq \sup_k |\alpha_k| \sum_{k=0}^{\infty} |z|^k = \frac{\sup_k |\alpha_k|}{1 - |z|}. \tag{5.42}$$

We claim that

$$|\alpha_k| \leq 4 \, \max \left\{ \frac{1}{7.2^{m/2}}, \frac{2}{m} \right\}, \qquad \forall k = 0, 1, ...,$$

where $\mathbf{x} = (\alpha_k)_{k \geq 0} \in B_1(0) \cap \ker \Phi$. If the latter inequalities hold, then, in view of (5.42), the proof of the theorem will have been completed.

Now, we estimate $|\alpha_k|$. In fact, $\mathbf{x} = (\alpha_k)_{k \geq 0}$ is in $\ker \Phi$ iff the following system of m linear equations is satisfied

$$\begin{cases} \alpha_0 + \alpha_1 z_1 + \ ... \ + \alpha_{m-1} z_1^{m-1} = -\sum_{n=0}^{\infty} \alpha_{m+n} z_1^{m+n} \equiv \psi_{m1} \\ \vdots \qquad \vdots \qquad \vdots \qquad\qquad\qquad \vdots \\ \alpha_0 + \alpha_1 z_m + \ ... \ + \alpha_{m-1} z_m^{m-1} = -\sum_{n=0}^{\infty} \alpha_{m+n} z_m^{m+n} \equiv \psi_{mm} \end{cases} \tag{5.43}$$

Using Cramer's rule, we can in view of (5.43) write

$$\alpha_k = D_m^{-1}(\xi) \begin{vmatrix} 1 & ... & z_1^{k-1} & \psi_{m1} & z_1^{k+1} & ... & z_1^{m-1} \\ \vdots & & \vdots & \vdots & \vdots & & \vdots \\ 1 & ... & z_m^{k-1} & \psi_{mm} & z_m^{k+1} & ... & z_m^{m-1} \end{vmatrix}$$

where $k = 0, 1, .., m-1$ and $D_m(\xi)$ is the Vandermonde determinant as a function of $\xi = (z_1, .., z_m)$, i.e.,

$$D_m(\xi) = \begin{vmatrix} 1 & z_1 & z_1^2 & ... & z_1^{m-1} \\ 1 & z_2 & z_2^2 & ... & z_2^{m-1} \\ \vdots & \vdots & \vdots & & \vdots \\ 1 & z_m & z_m^2 & ... & z_m^{m-1} \end{vmatrix} \tag{5.44}$$

From (5.43), (5.44) one has

$$\alpha_k = -\sum_{l=0}^{\infty} \alpha_{m+l} D_m(l, k, \xi) D_m^{-1}(\xi) \tag{5.45}$$

where

$$D_m(l, k, \xi) = \begin{vmatrix} 1 & z_1 & \cdots & z_1^{k-1} & z_1^{m+l} & z_1^{k+1} & \cdots & z_1^{m-1} \\ 1 & z_2 & \cdots & z_2^{k-1} & z_2^{m+l} & z_2^{k+1} & \cdots & z_2^{m-1} \\ \vdots & \vdots & & \vdots & \vdots & \vdots & & \vdots \\ 1 & z_m & \cdots & z_m^{k-1} & z_m^{m+l} & z_m^{k+1} & \cdots & z_m^{m-1} \end{vmatrix}$$

$l = 0, 1, 2, .., \ k = 0, 1, .., m - 1.$

We claim that $D_m(l, k, \xi) D_m^{-1}(\xi)$ is a homogeneous polynomial in $\xi = (z_1, z_2, .., z_m)$. We first set some notations. For $m, M = 1, 2, .., \ k = 1, 2, .., m - 1, \ \beta = (\beta_1, .., \beta_m), \ \xi = (z_1, z_2, .., z_m)$, we put

$$|\beta| = \beta_1 + ... + \beta_m, \quad \xi^\beta = z_1^{\beta_1} .. z_m^{\beta_m},$$

and

$$A(m, M) = \{\beta = (\beta_1, .., \beta_m) \in (\mathbf{Z}^+)^m : \ |\beta| = M\},$$
$$B(m, M, k) = \{\beta \in A(m, M) : \text{ there are } i_1, .., i_{m-k}$$
$$\text{in } \overline{1, m} \text{ such that } \beta_{i_j} \geq 1, \ j = 1, 2, .., m - k\}$$

where $\mathbf{Z}^+ = \{0, 1, 2, ..\}$. Now we prove that

$$D_m(l, k, \xi) D_m^{-1}(\xi) = \sum_{\beta \in B(m, l+m-k, k)} C_{\beta, l, k} \xi^\beta \tag{5.46}$$

where the coefficients $C_{\beta, l, k}$ satisfy

$$|C_{\beta, l, k}| \leq \binom{m}{k}. \tag{5.47}$$

We prove (5.46), (5.47) by induction on m.

Let $m = 2$. Then $\xi = (z_1, z_2), \ k = 0$ or $k = 1$.

If $k = 0$, we have

$$D_m(l, 0, \xi) D_m^{-1}(\xi) = -z_1 z_2 \sum_{j=0}^{l} z_1^{l-j} z_2^j$$

$$= -\sum_{j=0}^{l} z_1^{l-j+1} z_2^{j+1}.$$

Since

$$(l - j + 1, j + 1) \in B(2, l + 2 - 0, 0), \qquad j \in \overline{0, l},$$

we infer that (5.46), (5.47) hold for $m = 2, \ k = 0$.

Now, for $k = 1$, we have

$$D_m(l, 1, \xi) D_m^{-1}(\xi) = \sum_{j=0}^{l+1} z_1^{l-j+1} z_2^j.$$

Since

$$(l - j + 1, j) \in B(2, l + 2 - 1, 1), \qquad j \in \overline{0, l+1},$$

we infer that (5.46), (5.47) hold in this case.

Now, assuming that (5.46), (5.47) hold for $m = s$ $(s \geq 2)$ we shall prove that it also holds for $m = s + 1$.

For $\xi = (z_1, .., z_{s+1})$ we put $t = (t_2, .., t_{s+1})$ with

$$t_j = z_j z_1^{-1}, \qquad j = 2, 3, .., s + 1.$$

We can write

$$D_{s+1}(l, k, \xi) = z_1^{\frac{s(s+1)}{2} + s + l + 1 - k} \times$$

$$\times \begin{vmatrix} 1 & 1 & \ldots & 1 & 1 & 1 & \ldots & 1 \\ 1 & t_2 & \ldots & t_2^{k-1} & t_2^{s+l+1} & t_2^{k+1} & \ldots & t_2^s \\ \vdots & \vdots & & \vdots & \vdots & \vdots & & \vdots \\ 1 & t_{s+1} & \ldots & t_{s+1}^{k-1} & t_{s+1}^{s+l+1} & t_{s+1}^{k+1} & \ldots & t_{s+1}^s \end{vmatrix}$$

$$= z_1^{\frac{s(s+1)}{2} + s + l + 1 - k} \times$$

$$\times \begin{vmatrix} 1 & 1 & \ldots & 1 & 1 & 1 & \ldots & 1 \\ 0 & t_2 - 1 & \ldots & t_2^{k-1} - 1 & t_2^{s+l+1} - 1 & t_2^{k+1} - 1 & \ldots & t_2^s - 1 \\ \vdots & \vdots & & \vdots & \vdots & \vdots & & \vdots \\ 0 & t_{s+1} - 1 & \ldots & t_{s+1}^{k-1} - 1 & t_{s+1}^{s+l+1} - 1 & t_{s+1}^{k+1} - 1 & \ldots & t_{s+1}^s - 1 \end{vmatrix}$$

After some standard computations, we get

$$D_{s+1}(l, k, \xi) =$$

$$= z_1^{\frac{s(s+1)}{2} + s + l + 1 - k} \prod_{j=2}^{s+1} (t_j - 1) \times$$

$$\times \begin{vmatrix} 1 & t_2 & \ldots & t_2^{k-2} & \sum_{j=0}^l t_2^{s+j} & t_2^{k-1} + t_2^k & t_2^{k+1} & \ldots & t_2^{s-1} \\ \vdots & \vdots & & \vdots & \vdots & \vdots & \vdots & & \vdots \\ 1 & t_{s+1} & \ldots & t_{s+1}^{k-2} & \sum_{j=0}^l t_{s+1}^{s+j} & t_{s+1}^{k-1} + t_{s+1}^k & t_{s+1}^{k+1} & \ldots & t_{s+1}^{s-1} \end{vmatrix}$$

$$= z_1^{\frac{s(s+1)}{2} + s + l + 1 - k} \prod_{j=2}^{s+1} (t_j - 1) \times$$

$$\times \sum_{j=0}^{l} (-D_s(j, k, t) + D_s(j, k - 1, t))$$

$$= z_1^{\frac{s(s+1)}{2} + s + l + 1 - k} \prod_{j=2}^{s+1} (t_j - 1) \times$$

$$\times (D_s(t) \{ D_s(l, k - 1, t) D_s^{-1}(t) - D_s(0, k, t) D_s^{-1}(t) \}$$

$$+ D_s(t) \sum_{j=0}^{l-1} (D_s(j, k - 1, t) - D_s(j + 1, k, t)) D_s^{-1}(t))$$

$$\text{for } 3 \leq k \leq s - 1. \tag{5.48}$$

We have

$$D_s(t)z_1^{\frac{s(s+1)}{2}}\prod_{j=2}^{s+1}(t_j-1) = D_{s+1}(\xi).$$

Hence, relation (5.48) gives

$$D_{s+1}(l,k,\xi)D_{s+1}^{-1}(\xi) = z_1^{s+l+1-k}\times$$
$$\times\left(\{D_s(l,k-1,t)D_s^{-1}(t) - D_s(0,k,t)D_s^{-1}(t)\} + \right.$$
$$\left. + \sum_{j=0}^{l-1}\left(D_s(j,k-1,t) - D_s(j+1,k,t)\right)D_s^{-1}(t)\right)$$
$$\text{for } 3 \le k \le s-1. \tag{5.49}$$

Since (5.46), (5.47) hold for $m = s$ we get in view of (5.49) and the relation

$$B(s,j+s+1-k,k-1) \subset B(s,j+s+1-k,k)$$

that

$$D_{s+1}(l,k,\xi)D_{s+1}^{-1}(\xi) = z_1^{s+l+1-k}\times$$

$$\times\left\{\sum_{\tilde\beta\in B(s,s+l+1-k,k-1)}C_{\tilde\beta,l,k-1}t^{\tilde\beta}+\right.$$

$$+\sum_{j=0}^{l-1}\left(\sum_{\tilde\beta\in B(s,s+j+1-k,k)}(C_{\tilde\beta,j,k-1}-C_{\tilde\beta,j+1,k})t^{\tilde\beta}\right)+$$

$$\left.+\sum_{\tilde\beta\in B(s,s-k,k)}C_{\tilde\beta,0,k}t^{\tilde\beta}\right\} \tag{5.50}$$

where $3 \le k \le s-1$ and

$$|C_{\tilde\beta,j,k-1}| \le \binom{s}{k-1}, \qquad j = 0,1,..,l-1, \tag{5.51}$$

$$|C_{\tilde\beta,j,k}| \le \binom{s}{k}, \qquad j = 0,1,..,l. \tag{5.52}$$

Recall that $t = (t_2,..,t_{s+1}) = z_1^{-1}(z_2,..,z_{s+1})$. Thus, relation (5.24) implies

$$D_{s+1}(l,k,\xi)D_{s+1}^{-1}(\xi) =$$

$$=\sum_{\tilde\beta\in B(s,s+l+1-k,k-1)}C_{\tilde\beta,l,k-1}\xi^{(0,\tilde\beta)}+$$

$$+\sum_{j=0}^{l-1}\left(\sum_{\tilde\beta\in B(s,s+j+1-k,k)}(C_{\tilde\beta,j,k-1}-C_{\tilde\beta,j+1,k})\xi^{(l-j,\tilde\beta)}\right)+$$

$$+\sum_{\tilde\beta\in B(s,s-k,k)}C_{\tilde\beta,0,k}\xi^{(l+1,\tilde\beta)}. \tag{5.53}$$

In view of (5.51), (5.52) and the identity

$$\binom{s}{k-1} + \binom{s}{k} = \binom{s+1}{k}$$

it follows that all the coefficients of $D_{s+1}(l,k,\xi)D_{s+1}^{-1}(\xi)$ satisfy (5.47) for $m = s+1$, $3 \le k \le s-1$. On the other hand, we have

$$(0,\tilde{\beta}) \in B(s+1,s+l+1-k,k) \text{ for}$$
$$\tilde{\beta} \in B(s,s+l+1-k,k-1), \tag{5.54}$$
$$(l-j,\tilde{\beta}) \in B(s,s+l+1-k,k) \text{ for}$$
$$j \in \overline{0,l-1}, \ \tilde{\beta} \in B(s,s+j+1-k,k). \tag{5.55}$$
$$(l+1,\tilde{\beta}) \in B(s+1,s+l+1-k,k) \text{ for}$$
$$\tilde{\beta} \in B(s,s-k,k). \tag{5.56}$$

In view of (5.54)-(5.56), the relation (5.46) holds for $m = s+1$, $0 \le k \le s-1$.

Similarly as for (5.48), we shall get expansions for $D_{s+1}(l,j,\xi) \ D_{s+1}^{-1}(\xi)$ ($j = 0,1,2$), $D_{s+1}(l,s,\xi)D_{s+1}^{-1}(\xi)$. In fact, these expansions are simpler than (5.48). Similarly as for the case $3 \le k \le s-1$ we also get (5.46), (5.47) for $k = 0,1,2$ and $k = s$. Hence, (5.46), (5.47) hold for $m = s+1$, $0 \le k \le s$.

By the induction principle, (5.46), (5.47) hold for every $m \ge 2$.

We now turn to relation (5.45). Using (5.46), (5.47), one gets

$$|\alpha_k| \le \sum_{l=0}^{\infty} |\alpha_{m+l}| C_m^k \alpha^{m+l-k} card B(m,l+m-k,k),$$
$$k = 1,2,..,m, \tag{5.57}$$

where $card B(m,l+m-k,k)$ is the number of the elements of $B(m,l+m-k,k)$. We propose to derive an estimate of this number. We claim that

$$card B(m,l+m-k,k) \le \binom{m}{k}\binom{l+m-1}{m-1}. \tag{5.58}$$

Put

$$A_1(m,k) = \{\beta = (\beta_1,..,\beta_m) : \text{ there are } i_1,1_2,..,i_{m-k}$$
$$\text{in } \overline{1,m} \text{ such that } \beta_{i_j} = 1 \ \forall j = \overline{1,m-k}$$
$$\text{and } \beta_i = 0 \text{ if } i \in \{i_1,..,i_{m-k}\}\}.$$

Consider the mapping

$$\Psi : \ A(m,l) \times A_1(m,k) \longrightarrow B(m,l+m-k,k),$$

$$\Psi(\beta,\gamma) = \beta + \gamma, \quad (\beta,\gamma) \in A(m,l) \times A_1(m,k).$$

We prove that Ψ is surjective. Letting $\zeta = (\zeta_1,..,\zeta_m)$ be in $B(m,l+m-k,k)$, one has $|\zeta| = m+l-k$ and there are $i_1,..,i_{m-k}$ such that

$$\zeta_{i_j} \geq 1, \qquad j = 1, 2, .., m - k.$$

Put

$$\gamma = (0, .., 0, \underbrace{1}_{i_1-th}, 0, .., 0, \underbrace{1}_{i_j-th}, 0, .., 0, \underbrace{1}_{i_{m-k}-th}, 0, .., 0) \in A_1(m, k).$$

and

$$\beta = \zeta - \gamma.$$

Then $\beta \in A(m, l)$. Moreover, $\Psi(\beta, \gamma) = \zeta$. Hence Ψ is surjective. Thus we have

$$card B(m, m + l - k, k) \leq card A(m, l).card A_1(m, k). \tag{5.59}$$

The number of the elements of $A_1(m, k)$ is the number of $(m - k)$-element subsets of an n-element set. Hence

$$card A_1(m, k) = \binom{m}{k}. \tag{5.60}$$

The number of the elements of $A(m, l)$ is the number of solutions of the equation

$$\beta_1 + .. + \beta_m = l, \qquad \beta_i \in \overline{0, l}, \ i = 1, 2, .., m.$$

We know that the number of solutions of this equation is $\binom{m+l-1}{m-1}$ (see, e.g., [Di], chap. 3, page 114). Hence

$$card A(m, l) = \binom{m + l - 1}{m - 1}. \tag{5.61}$$

Combining (5.59)-(5.61) gives (5.58). Substituting (5.58) into (5.57) we get

$$|\alpha_k| \leq \sum_{l=0}^{\infty} |\alpha_{m+l}| \binom{m}{k}^2 \binom{m + l - 1}{m - 1} \alpha^{m+l-k}. \tag{5.62}$$

In view of the inequalities

$$\binom{m}{k} \leq 2^m, \quad \binom{m + l - 1}{m - 1} \leq 2^{m+l-1},$$

inequality (5.62) gives

$$|\alpha_k| \leq \sum_{l=0}^{\infty} |\alpha_{m+l}| 2^{3m+l-1} \alpha^{m+l-k}, \quad 0 \leq k \leq m - 1. \tag{5.63}$$

Since $\mathbf{x} = (\alpha_k)_{k \geq 0} \in B_1(0)$, one has

$$|\alpha_0|^2 + \sum_{k=1}^{\infty} k^2 |\alpha_k|^2 \leq 1. \tag{5.64}$$

In view of (5.64), the inequality (5.63) for $0 < \alpha \leq 1/16$ implies

$$|\alpha_k| \leq \sum_{l=0}^{\infty} 2^{3m+l-1} \cdot 2^{-4m-4l+4k}$$

$$= \sum_{l=0}^{\infty} 2^{-3l-m-1+4k}$$

$$= 2^{-m-1+4k} \frac{1}{1-1/8}$$

$$= \frac{4}{7} 2^{-m+4k}, \qquad \forall \mathbf{x} = (\alpha_k)_{k \geq 0} \in B_1(0) \cap \ker \Phi.$$

Hence, for $0 \leq k \leq m/8$ one has

$$|\alpha_k| \leq \frac{4}{7} 2^{-m/2}, \qquad \forall \mathbf{x} = (\alpha_k)_{k \geq 0} \in B_1(0) \cap \ker \Phi. \tag{5.65}$$

From (5.64), one has

$$|\alpha_k| \leq k^{-1}, \qquad k = 1, 2, .., \forall \mathbf{x} = (\alpha_k)_{k \geq 0} \in B_1(0).$$

Hence, if $k \geq m/8$ then

$$|\alpha_k| \leq 8/m, \qquad \forall \mathbf{x} = (\alpha_k)_{k \geq 0} \in B_1(0). \tag{5.66}$$

It follows from (5.65), (5.66) that

$$|\alpha_k| \leq \max\left\{ \frac{4}{7} 2^{-m/2}, \frac{8}{m} \right\}, \quad \forall \mathbf{x} = (\alpha_k)_{k \geq 0} \in B_1(0) \cap \ker \Phi. \tag{5.67}$$

Finally, we estimate

$$\sup_{\mathbf{x} \in B_1(0) \cap \ker \Phi} |\varphi_z(\mathbf{x})|.$$

By the definition, one has

$$\varphi_z(\mathbf{x}) = \sum_{k=0}^{\infty} \alpha_k z^k.$$

Hence

$$|\varphi_z(\mathbf{x})| \leq \sup_k |\alpha_k| \sum_{k=0}^{\infty} |z|^k = \frac{\sup_k |\alpha_k|}{1-|z|}. \tag{5.68}$$

If $\mathbf{x} \in B_1(0) \cap \ker \Phi$, then (5.67) holds.

From (5.67), (5.68), we get

$$\sup_{\mathbf{x} \in B_1(0) \cap \ker \Phi} |\varphi_z(\mathbf{x})| \leq \frac{4}{1-|z|} \max\left\{ \frac{1}{7.2^{m/2}}, \frac{2}{m} \right\}.$$

By Proposition 5.2, the latter inequality implies

$$\sup_{\mathbf{x} \in B_{C_0}(0)} \left| \varphi_z(\mathbf{x}) - \sum_{j=1}^{m} b_j \sum_{k=0}^{\infty} \alpha_k z_j^k \right| = C_0 \sup_{\mathbf{x} \in B_1(0) \cap \ker \Phi} |\varphi_z(\mathbf{x})|$$

$$\leq \frac{4C_0}{1 - |z|} \max \left\{ \frac{1}{7.2^{m/2}}, \frac{2}{m} \right\}, \quad \forall z \in U.$$

This is the first inequality in Theorem 5.3. Now, for $u \in H^2(U)$ satisfying $u' \in H^2(U)$, the function u can be represented by the series

$$u(z) = \sum_{k=0}^{\infty} a_k z^k$$

with

$$\sum_{k=0}^{\infty} |a_k|^2 + \sum_{k=0}^{\infty} k^2 |a_k|^2 = \|u\|_{H^2}^2 + \|u'\|_{H^2}^2 < \infty.$$

Hence $\mathbf{x}_0 \equiv (a_k)_{k \geq 0} \in B_{C_0}(0)$ with

$$C_0 \equiv C_0' = \|u\|_{H^2} + \|u'\|_{H^2}.$$

(Here, we recall,

$$B_{C_0}(0) = \left\{ (\alpha_k)_{k \geq 0} : \left(|\alpha_0|^2 + \sum_{k=1}^{\infty} k^2 |\alpha_k|^2 \right)^{1/2} \leq C_0 \right\})$$

We also have

$$\varphi_z(\mathbf{x}_0) = \sum_{k=0}^{\infty} a_k z^k = u(z), \tag{5.69}$$

$$\sum_{k=0}^{\infty} a_k z_j^k = u(z_j) = \mu_j. \tag{5.70}$$

Hence

$$\left| u(z) - \sum_{j=1}^{m} b_{mj}(z) \mu_j \right| = \left| \varphi_z(\mathbf{x}_0) - \sum_{j=1}^{m} b_{mj}(z) \mu_j \right|$$

$$\leq \sup_{\mathbf{x} \in B_{C_0}(0)} \left| \varphi_z(\mathbf{x}) - \sum_{j=1}^{m} b_{mj}(z) \sum_{k=0}^{\infty} \alpha_k z_j^k \right|. \tag{5.71}$$

From (5.68), (5.71), we obtain

$$\left| u(z) - \sum_{j=1}^{m} b_{mj}(z) \mu_j \right| \leq \frac{4C_0'}{1 - |z|} \max \left\{ \frac{1}{7.2^{m/2}}, \frac{2}{m} \right\}. \tag{5.72}$$

Now, if $z \in K$ (a compact subset of U) then

$$\frac{1}{1 - |z|} \leq \frac{1}{\text{dist}\,(K, \partial U)}.$$ (5.73)

From (5.72), (5.73) we get

$$\left| u(z) - \sum_{j=1}^{m} b_{mj}(z)\mu_j \right| \leq \frac{4C_0'}{\text{dist}\,(K, \partial U)} \max\left\{ \frac{1}{7.2^{m/2}}, \frac{2}{m} \right\}, \quad \forall z \in K.$$

Hence

$$\sup_{z \in K} \left| u(z) - \sum_{j=1}^{m} b_{mj}(z)\mu_j \right| \leq \frac{4C_0'}{\text{dist}\,(K, \partial U)} \max\left\{ \frac{1}{7.2^{m/2}}, \frac{2}{m} \right\}.$$

This completes the proof of Theorem 5.3.

5.3 Cardinal series representation and approximation: reformulation of moment problems

In this section, we derive some theorems on cardinal series representation and approximation by cardinal series. Although these results of independent interest, their derivation has been motivated by the problem of reformulation of moment problems as infinite systems of linear equations or approximation of moment problems by infinite systems of linear equations.

We commence with a cardinal series representation theorem for functions in $L^2(\mathbf{R}^2)$ the Fourier transforms of which have compact supports. A one-dimensional version was proved in [St] (see Theorem 5.5 below). For the two-dimensional case, we follow the approach of [St] (loc. cit.). An illustration is given of the application to the reformulation of a moment problem on $\mathbf{R}$ as an infinite system of linear algebraic equations in infinitely many unknowns. The section closes with two Sinc approximation theorems on $\mathbf{R}$ and $\mathbf{R}^2$ respectively. These approximation theorems are, to our knowledge, new.

5.3.1 Two-dimensional Sinc theory

Let f be a function in $L^2(\mathbf{R}^2)$. Define the Fourier transform of f by

$$\hat{f}(s, t) = \int_{-\infty}^{\infty} \int_{-\infty}^{\infty} f(x, y)e^{ixs + iyt}\,dx\,dy.$$ (5.74)

Suppose the support of $\hat{f}$ is compact,

$$\text{supp}\,\hat{f} \subset [-\pi/h, \pi/h] \times [-\pi/k, \pi/k].$$

We propose to represent f by a double series of Sinc functions, analogous to the one-dimensional case.

Putting

$$F(s,t) = \hat{f}(s,t)$$

we have by the inversion formula

$$f(x,y) = \frac{1}{(2\pi)^2} \int_{-\pi/h}^{\pi/h} \int_{-\pi/k}^{\pi/k} F(s,t)e^{-isx-ity}\,dsdt. \qquad (5.75)$$

Now, we have

$$F(s,t) \sim \sum_{m,n} c_{mn} e^{imhs+inkt} \qquad (5.76)$$

where

$$\begin{aligned} c_{mn} &= \frac{hk}{(2\pi)^2} \int_{-\pi/h}^{\pi/h} \int_{-\pi/k}^{\pi/k} F(s,t)e^{-imhs-inkt}\,dsdt \\ &= hkf(mh,nk). \end{aligned} \qquad (5.77)$$

Letting $|m| \leq |n|$ in the double Fourier series (5.76), we have from (5.76)

$$F(s,t) = hk \sum_{n=-\infty}^{\infty} \sum_{|m|\leq|n|} f(mh,nk)e^{imhs+inkt}, \qquad (5.78)$$

that is,

$$\begin{aligned} F(s,t) &= hk \sum_{n=-\infty}^{\infty} \sum_{|m|\leq|n|} f(mh,nk)e^{imhs+inkt} \\ &\quad for\ (s,t) \in [-\pi/h, \pi/h] \times [-\pi/k, \pi/k], \\ &= 0\ \ for\ (s,t) \notin [-\pi/h, \pi/h] \times [-\pi/k, \pi/k], \end{aligned} \qquad (5.79)$$

where the Fourier series in (5.76) converges in $L^2(\mathbf{R}^2)$ (and also a.e.).

As in [St], p. 91, we define

$$S(p,d)(z) = \frac{\sin[\pi(z-pd)/d]}{\pi(z-pd)/d}, \quad p \in \mathbf{Z},\ d > 0. \qquad (5.80)$$

Substituting the right hand side of (5.79) into the right hand side of (5.75) and integrating termwise, we get

$$f(x,y) = \sum_{n=-\infty}^{\infty} \sum_{|m|\leq|n|} f(mh,nk)S(m,h)(x)S(n,k)(y). \qquad (5.81)$$

In the derivation of (5.81), we have used the relation

$$\frac{1}{2\pi} \int_{-\pi/d}^{\pi/d} de^{ipdt-izt}\,dt = S(p,d)(z), \quad p \in \mathbf{Z},\ d > 0. \qquad (5.82)$$

The normalized Sinc functions $h^{-1/2}S(m,h)(x)k^{-1/2}S(n,k)(y)$ form a complete orthonormal sequence in $L^2(\mathbf{R}^2)$. Indeed, we have

$$
\int_{R^2} S(m,h)(x)S(n,k)(y)S(l,h)(x)S(j,k)(y)dxdy =
$$
$$
= \frac{h^2 k^2}{(2\pi)^2} \int_{-\pi/h}^{\pi/h} \int_{-\pi/k}^{\pi/k} e^{i(m-l)ht+i(n-j)kx} dt dx
$$
$$
= \begin{cases} hk & \text{if } m = l, n = j, \\ 0 & \text{otherwise.} \end{cases}
$$

Summarizing, we have

Theorem 5.4. *Let $h, k > 0$. The family $\{\varphi_{mn}\}$ where*

$$
\varphi_{mn}(x,y) = h^{-1/2}k^{-1/2}S(m,h)(x)S(n,k)(y)
$$

is a complete orthonormal sequence in the space $W(\frac{\pi}{h}, \frac{\pi}{k})$ of $L^2(\mathbf{R}^2)$-functions, the Fourier transforms of which have supports contained in $[-\pi/h, \pi/h] \times [-\pi/k, \pi/k]$

In the one-dimensional case, we have

Theorem 5.5. *Let $h > 0$. Then the family of functions $\{h^{-1/2}S(n,h)\}$ is a complete orthonormal function in the space $W(\frac{\pi}{h})$ of $L^2(\mathbf{R})$-functions, the Fourier transforms of which have supports contained in $[-\pi/h, \pi/h]$.*

Consider now the moment problem

$$
\int_{\mathbf{R}} u(x)g_k(x)dx = \mu_k, \qquad k = 1, 2, ..., \tag{5.83}
$$

where (g_k) is a given sequence of functions in $W(\pi/h)$.

Assuming $u \in W(\pi/h)$, we expand $u(x)$ and each $g_k(x)$ into cardinal series

$$
u(x) = \sum_{n \in \mathbf{Z}} x_n S(n,h)(x),
$$
$$
g_k(x) = \sum_{n \in \mathbf{Z}} a_{nk} S(n,h)(x).
$$

Using the orthonormality of the $h^{-1/2}S(n,h)$'s, we arrive at an infinite set of linear equations in infinitely many unknowns

$$
\sum_{n \in \mathbf{Z}} a_{nk} x_n = \mu_k, \qquad k = 1, 2, ... \tag{5.84}
$$

This is, in general, an ill-posed problem. A regularization of (5.84) can be achieved by using methods presented in previous chapters. Note that a moment problem on $\mathbf{R}^2$ can, under appropriate hypotheses, be reduced, using two-dimensional cardinal series representations, to a linear system of infinitely many equations for which regularization methods of the previous chapters can be applied. We do not go into details.

5.3.2 Approximation theorems

We close this chapter with two approximation theorems that will be useful in our regularization of various problems in subsequent chapters. Let f be a function in $L^2(\mathbf{R})$ such that for a positive number h, $f(x)$ is known at equidistant points $x = nh$, $n \in \mathbf{Z}$. We propose to approximate f by a function in $W(\frac{\pi}{h})$, the space introduced in Theorem 5.5. We note first that f has to be (at least) continuous in order for $f(nh)$ to be defined. We need other conditions on f, namely

$$(1 + \omega^2)^j \hat{f} \in L^2(\mathbf{R}) \cap L^\infty(\mathbf{R}), \quad , \tag{5.85}$$

$$\hat{f}' \text{ exists and}$$

$$(1 + \omega^2)^j \hat{f}' \in L^2(\mathbf{R}), \tag{5.86}$$

where $j \geq 1$ is an integer.

We put

$$\tilde{f} = \sum_{n \in \mathbf{Z}} f(nh) S(n, h)(x). \tag{5.87}$$

We claim that if f satisfies (5.85), (5.86), then this series converges in $L^2(\mathbf{R})$. In fact, integrating by parts we have

$$f(x) = \frac{1}{2\pi} \int_{-\infty}^{\infty} \hat{f}(\omega) e^{-ix\omega} d\omega$$

$$= \frac{1}{2\pi} \int_{-\infty}^{\infty} \frac{\hat{f}'(\omega) e^{-i\omega x}}{ix} d\omega.$$

Hence

$$|f(nh)| \leq \frac{1}{2\pi} \int_{-\infty}^{\infty} \frac{|\hat{f}'(\omega)|}{|n|h} d\omega$$

$$\leq \frac{1}{2\pi|n|h} \int_{-\infty}^{\infty} (1 + \omega^2)^{-1/2} (1 + \omega^2)^{1/2} |\hat{f}'(\omega)| d\omega$$

$$\leq \frac{1}{2\pi|n|h} \left(\int_{-\infty}^{\infty} (1 + \omega^2)^{-1} d\omega \right)^{1/2} \times$$

$$\times \left(\int_{-\infty}^{\infty} (1 + \omega^2) |\hat{f}'(\omega)|^2 d\omega \right)^{1/2}$$

$$\leq \frac{1}{2\sqrt{\pi}|n|h} \|(1 + \omega^2)^{1/2} \hat{f}'\|_{L^2(\mathbf{R})}. \tag{5.88}$$

By (5.88), $(f(nh))_{n \in \mathbf{Z}} \in l^2(\mathbf{Z})$ where

$$l^2(\mathbf{Z}) = \left\{ (\alpha_n)_{n \in \mathbf{Z}} : \sum_{n \in \mathbf{Z}} |\alpha_n|^2 < \infty \right\}.$$

Hence the series in (5.87) converges in $L^2(\mathbf{R})$.

Theorem 5.6. *Let $h \in (0,1)$ and let $j \geq 1$ be an integer. Suppose that f satisfies (5.85), (5.86). Then there exists a C depending only on f, j such that*

$$\|f - \tilde{f}\|_{L^2(\mathbf{R})} \leq Ch^{2j-1}$$

where $\tilde{f}$ is defined in (5.87).

Proof.

Put

$$f_h(x) = \frac{1}{2\pi} \int_{-\pi/h}^{\pi/h} \hat{f}(\omega)e^{-i\omega x} d\omega. \tag{5.89}$$

Then

$$\hat{f}_h(\omega) = \begin{cases} \hat{f}(\omega), & \omega \in (-\pi/h, \pi/h), \\ 0, & \omega \notin (-\pi/h, \pi/h). \end{cases}$$

i.e. $f_h \in W(\pi/h)$. We also note that

$$f(x) = \frac{1}{2\pi} \int_{-\infty}^{\infty} \hat{f}(\omega)e^{-i\omega x} d\omega. \tag{5.90}$$

We have in view of (5.89), (5.90)

$$\|f - f_h\|_{L^2(\mathbf{R})}^2 = \frac{1}{2\pi}\|\hat{f} - \hat{f}_h\|_{L^2(\mathbf{R})}^2$$

$$= \frac{1}{2\pi} \int_{|\omega|>\pi/h} |\hat{f}(\omega)|^2 d\omega.$$

Using (5.85) one has

$$\|f - f_h\|_{L^2(\mathbf{R})} = \frac{1}{\sqrt{2\pi}} \left(\int_{|\omega|>\pi/h} (1+\omega^2)^{-2j}[(1+\omega^2)^j|\hat{f}(\omega)|]^2 d\omega \right)^{1/2}$$

$$\leq \frac{h^{2j}}{\sqrt{2\pi}2^{2j+1/2}}\|(1+\omega^2)^j|\hat{f}| \,\|_{L^2(\mathbf{R})}. \tag{5.91}$$

On the other hand, since $f_h \in W(\frac{\pi}{h})$, we can expand f_h in a cardinal series

$$f_h(x) = \sum_{n\in\mathbf{Z}} f_h(nh)S(n,h)(x). \tag{5.92}$$

Hence (5.87), (5.92) give:

$$\tilde{f}(x) - f_h(x) = \sum_{n\in\mathbf{Z}} (f(nh) - f_h(nh))S(n,h)(x).$$

By the orthonormality of $\{h^{-1/2}S(n,h)\}$ in $L^2(\mathbf{R})$, this equality gives

$$\|\tilde{f} - f_h\|_{L^2(\mathbf{R})}^2 = h \sum_{n \in \mathbf{Z}} |f(nh) - f_h(nh)|^2. \tag{5.93}$$

But we have in view of (5.90), (5.89)

$$f(nh) - f_h(nh) = \frac{1}{2\pi} \int_{|w| > \pi/h} \hat{f}(\omega) e^{-inh\omega} d\omega.$$

Integrating here by parts the right hand side gives

$$|f(nh) - f_h(nh)| \leq \frac{1}{2\pi h |n|} \left(|\hat{f}(-\pi/h)| + |\hat{f}(\pi/h)| \right)$$

$$+ \frac{1}{2\pi h |n|} \left| \int_{|w| > \pi/h} \hat{f}'(\omega) e^{-inh\omega} d\omega \right|. \tag{5.94}$$

In view of (5.85), this gives

$$\frac{1}{2\pi h |n|} \left(|\hat{f}(-\pi/h)| + |\hat{f}(\pi/h)| \right) \leq \frac{Ch^{2j-1}}{|n|} \tag{5.95}$$

where C is a generic constant depending only on f, j.

For the third term of the right hand side of (5.94) we have

$$\frac{1}{2\pi h |n|} \left| \int_{|w| > \pi/h} \hat{f}(\omega) e^{-inh\omega} d\omega \right| \leq$$

$$\leq \frac{1}{2\pi h |n|} \int_{|w| > \pi/h} (1 + \omega^2)^{-j} (1 + \omega^2)^j |\hat{f}'(\omega)| d\omega$$

$$\leq \frac{1}{2\pi h |n|} \left(\int_{|w| > \pi/h} (1 + \omega^2)^{-2j} d\omega \right)^{1/2} \|(1 + \omega^2)^j \hat{f}'(\omega)\|_{L^2(\mathbf{R})}.$$

Hence

$$\frac{1}{2\pi h |n|} \left| \int_{|w| > \pi/h} \hat{f}(\omega) e^{-inh\omega} d\omega \right| \leq$$

$$\leq \frac{C}{|n| h} \left(2 \int_{\pi/h}^{\infty} \omega^{-4j} d\omega \right)^{1/2}$$

$$\leq \frac{C' h^{2j-3/2}}{|n|}. \tag{5.96}$$

where C' is also a generic constant depending only on f, j.

Combining (5.94)-(5.96) gives

$$|f(nh) - f_h(nh)| \leq \frac{Ch^{2j-3/2}}{|n|}. \tag{5.97}$$

Using (5.97), the equality (5.93) implies

$$\|\tilde{f} - f_h\|_{L^2(\mathbf{R})}^2 \leq 2hC(h^{2j-3/2})^2 \sum_{n=1}^{\infty} \frac{1}{n^2}$$

$$\leq C' h^{4j-2}. \tag{5.98}$$

From (5.91), (5.98) one has

$$\|\tilde{f} - f\|_{L^2(\mathbf{R})} \leq \|f - f_h\|_{L^2(\mathbf{R})} + \|\tilde{f} - f_h\|_{L^2(\mathbf{R})}$$
$$\leq C(h^{2j} + h^{2j-1})$$
$$\leq C' h^{2j-1}$$

where C' is a constant independent from h. This completes the proof of Theorem 5.6.

Now, we consider the problem of approximating a function $g : \mathbf{R}^2 \to \mathbf{R}$ satisfying

$$g(mh, nk) = \mu_{mn}, \qquad m, n = 0, \pm 1, \pm 2, \ldots \tag{5.99}$$

As in the one-dimensional case, we shall approximate g by functions in $W(\frac{\pi}{h}, \frac{\pi}{h})$. We assume $g \in C(\mathbf{R}^2)$ and

$$(1 + \omega^2 + \eta^2)^j \hat{g} \in L^2(\mathbf{R}^2) \cap L^{\infty}(\mathbf{R}^2), \tag{5.100}$$

$$(1 + \omega^2 + \eta^2)^j \frac{\partial \hat{g}}{\partial \eta}(\omega, .) \in L^{\infty}(\mathbf{R}, L^2(\mathbf{R})), \tag{5.101}$$

$$(1 + \omega^2 + \eta^2)^j \frac{\partial \hat{g}}{\partial \omega}(., \eta) \in L^{\infty}(\mathbf{R}, L^2(\mathbf{R})), \tag{5.102}$$

$$(1 + \omega^2 + \eta^2)^j \frac{\partial^2 \hat{g}}{\partial \omega \partial \eta} \in L^2(\mathbf{R}^2), \tag{5.103}$$

where $j \geq 2$ is an integer.

Theorem 5.7. *Let $j = 2, 3, \ldots$, $h \in (0, 1)$ and let g satisfy (5.100)-(5.103). Then there exists $C > 0$ depending only on j and g such that*

$$\|g - \tilde{g}\|_{L^2(\mathbf{R}^2)} \leq Ch^{j-3/2}$$

where

$$\tilde{g}(x, y) = \sum_{m,n=-\infty}^{\infty} \mu_{mn} S(m, h)(x) S(n, h)(y). \tag{5.104}$$

Proof
Put

$$g_h = \frac{1}{(2\pi)^2} \int_{-\pi/h}^{\pi/h} \int_{-\pi/h}^{\pi/h} \hat{g}(\omega, \eta) e^{-ix\omega - iy\eta} d\omega d\eta. \tag{5.105}$$

Then

$$\hat{g}_h(\omega, \eta) = \begin{cases} \hat{g}(\omega, \eta), & (\omega, \eta) \in [-\pi/h, \pi/h] \times [-\pi/h, \pi/h], \\ 0, & (\omega, \eta) \notin [-\pi/h, \pi/h] \times [-\pi/h, \pi/h], \end{cases}$$

i.e., $g_h \in W(\pi/h, \pi/h)$. We therefore have

$$\|\hat{g} - \hat{g}_h\|_{L^2(\mathbf{R}^2)}^2$$

$$\leq \int_{|w| > \pi/h} \int_{\mathbf{R}} |\hat{g}(\omega, \eta)|^2 d\eta d\omega + \int_{\mathbf{R}} \int_{|\eta| > \pi/h} |\hat{g}(\omega, \eta)|^2 d\eta d\omega$$

$$\leq \int_{|w| > \pi/h} \int_{\mathbf{R}} (1 + \omega^2 + \eta^2)^{-2j} (1 + \omega^2 + \eta^2)^{2j} |\hat{g}(\omega, \eta)|^2 d\eta d\omega +$$

$$+ \int_{\mathbf{R}} \int_{|\eta| > \pi/h} (1 + \omega^2 + \eta^2)^{-2j} (1 + \omega^2 + \eta^2)^{2j} |\hat{g}(\omega, \eta)|^2 d\omega d\eta$$

$$\leq 2 \left(\frac{\pi}{h} \right)^{-4j} \int_{\mathbf{R}^2} (1 + \omega^2 + \eta^2)^{2j} |\hat{g}(\omega, \eta)|^2 d\omega d\eta.$$

In view of (5.100), the latter inequality gives

$$\|\hat{g} - \hat{g}_h\|_{L^2(\mathbf{R}^2)} \leq C h^{2j}. \tag{5.106}$$

From (5.82), one has

$$g_h(x, y) = \sum_{-\infty}^{\infty} \sum_{|m| \leq |n|} g_h(mh, nh) S(m, h)(x) S(n, h)(y). \tag{5.107}$$

Subtracting (5.104) from (5.107), we get

$$g_h(x, y) - \tilde{g}(x, y) = \sum_{n=-\infty}^{\infty} \sum_{|m| \leq |n|} (g_h(mh, nh) - \mu_{mn}) \times$$
$$\times S(m, h)(x) S(n, h)(y).$$

Hence, by the orthonormality of $\{h^{-1/2} S(m, h)(x) h^{-1/2} S(n, h)(y)\}_{m,n}$,

$$\|g_h - \tilde{g}\|^2 \leq h^2 \sum_{m,n=0}^{\infty} |g_h(mh, nh) - g(mh, nh)|^2. \tag{5.108}$$

Now we have by (5.105)

$$g_h(mh, nh) = \frac{1}{(2\pi)^2} \int_{-\pi/h}^{\pi/h} \int_{-\pi/h}^{\pi/h} \hat{g}(\omega, \eta) e^{-imh\omega - inkn\eta} d\omega d\eta.$$

On the other hand

$$g(mh, nh) = \frac{1}{(2\pi)^2} \int_{\mathbf{R}^2} \hat{g}(\omega, \eta) e^{-imh\omega - inh\eta} d\omega d\eta.$$

Hence

$$g_h(mh, nh) - g(mh, nh) =$$

$$= \frac{1}{(2\pi)^2} \left(\int_{|\omega|<\pi/h} \int_{|\eta|>\pi/h} \right.$$

$$+ \left. \int_{|\omega|>\pi/h} \int_{\mathbf{R}} \right) \hat{g}(\omega, \eta) e^{-imh\omega - inh\eta} \, d\eta d\omega. \tag{5.109}$$

Now, integrating by parts we find

$$\left| \int_{|\omega|<\pi/h} \int_{|\eta|>\pi/h} \hat{g}(\omega, \eta) e^{-imh\omega - ink\eta} \, d\omega d\eta \right| \leq$$

$$\leq \frac{1}{|mn|h^2} \left(|\hat{g}(\frac{\pi}{h}, \frac{\pi}{h})| + |\hat{g}(\frac{\pi}{h}, -\frac{\pi}{h})| \right.$$

$$+ |\hat{g}(-\frac{\pi}{h}, \frac{\pi}{h})| + |\hat{g}(-\frac{\pi}{h}, -\frac{\pi}{h})| +$$

$$+ \int_{|\eta|>\pi/h} \left(\left| \frac{\partial \hat{g}}{\partial \eta}(\pi/h, \eta) \right| + \left| \frac{\partial \hat{g}}{\partial \eta}(-\pi/h, \eta) \right| \right) d\eta +$$

$$+ \int_{|\omega|<\pi/h} \left(\left| \frac{\partial \hat{g}}{\partial \omega}(\omega, \pi/h) \right| + \left| \frac{\partial \hat{g}}{\partial \omega}(\omega, -\pi/h) \right| \right) d\omega +$$

$$+ \left. \int_{|\omega|<\pi/h} \int_{|\eta|>\pi/h} \left| \frac{\partial^2 \hat{g}}{\partial \omega \partial \eta}(\omega, \eta) \right| d\eta d\omega \right). \tag{5.110}$$

From (5.100), one gets

$$|\hat{g}(\pm\pi/h, \pm\pi/h)| \leq C \left(1 + (\pi/h)^2 + (\pi/h)^2 \right)^{-j}$$

$$\leq C' h^{2j}. \tag{5.111}$$

From (5.101), it follows that

$$\int_{|\eta|>\pi/h} \left| \frac{\partial \hat{g}}{\partial \eta}(\pm\pi/h, \eta) \right| d\eta =$$

$$= \int_{|\eta|>\pi/h} \left(1 + (\pi/h)^2 + \eta^2 \right)^{-j} \left(1 + (\pi/h)^2 + \eta^2 \right)^j \left| \frac{\partial \hat{g}}{\partial \eta}(\pm\pi/h, \eta) \right| d\eta$$

$$\leq \left(\int_{|\eta|>\pi/h} \left(1 + (\pi/h)^2 + \eta^2 \right)^{-2j} d\eta \right)^{1/2} \times$$

$$\times \| \left(1 + (\pi/h)^2 + \eta^2 \right)^j \frac{\partial \hat{g}}{\partial \eta}(\pm\pi/h, \cdot) \|_{L^2(\mathbf{R}^2)}$$

$$\leq C \left(\int_{-\infty}^{\infty} \left((\pi/h)^2 + \eta^2 \right)^{-2j} d\eta \right)^{1/2}$$

$$= C \left(\frac{h^{4j-1}}{\pi^{4j-1}} \int_{-\infty}^{\infty} (1 + \eta_1)^{-2j} d\eta_1 \right)^{1/2}$$

$$\leq C' h^{2j-1/2} \tag{5.112}$$

where $\eta_1 = h\eta/\pi$.

From (5.102), the same estimates as for (5.112) give

$$\int_{|\omega|<\pi/h} \left| \frac{\partial \hat{g}}{\partial \omega}(\omega, -\pi/h) \right| d\omega \leq Ch^{2j-1/2}. \tag{5.113}$$

Now, from (5.103) we have

$$\int_{|\omega|<\pi/h} \int_{|\eta|>\pi/h} \left| \frac{\partial^2 \hat{g}}{\partial \omega \partial \eta}(\omega, \eta) \right| d\eta d\omega =$$

$$= \int_{|\omega|<\pi/h} \int_{|\eta|>\pi/h} \left(1 + \omega^2 + \eta^2\right)^{-j} \left(1 + \omega^2 + \eta^2\right)^{j} \left| \frac{\partial^2 \hat{g}}{\partial \omega \partial \eta}(\omega, \eta) \right| d\eta d\omega$$

$$\leq \left(\int_{|\omega|<\pi/h} \int_{|\eta|>\pi/h} \left(1 + \omega^2 + \eta^2\right)^{-2j} \right)^{1/2} \left\| \left(1 + \omega^2 + \eta^2\right)^{j} \frac{\partial^2 \hat{g}}{\partial \omega \partial \eta} \right\|_{L^2(\mathbf{R}^2)}$$

$$\leq \left(\int_{|\eta|>\pi/h} \left(\eta^2\right)^{-j} d\eta \int_{\mathbf{R}} \left(1 + \omega^2\right)^{-j} d\omega \right)^{1/2}$$

$$\leq C'h^{j-1/2}. \tag{5.114}$$

From (5.110)-(5.114), it follows that

$$\left| \int_{|\omega|<\pi/h} \int_{|\eta|>\pi/h} \hat{g}(\omega, \eta) e^{-imh\omega - inh\eta} d\eta d\omega \right| \leq \frac{Ch^{j-5/2}}{|mn|}. \tag{5.115}$$

Similarly, one has

$$\left| \int_{|\omega|>\pi/h} \int_{\mathbf{R}} \hat{g}(\omega, \eta) e^{-imh\omega - inh\eta} d\eta d\omega \right| \leq \frac{Ch^{j-5/2}}{|mn|}. \tag{5.116}$$

From (5.115), (5.116) we get in view of (5.109)

$$|g(mh, nh) - g_h(mh, nh)|^2 \leq \frac{Ch^{j-5/2}}{|mn|}.$$

By this inequality, the equation (5.108) implies

$$\|g_h - \tilde{g}\|_{L^2(\mathbf{R}^2)}^2 \leq Ch^2 \cdot h^{2j-5} \left(\sum_{n \in \mathbf{Z}} \frac{1}{n^2} \right) \left(\sum_{m \in \mathbf{Z}} \frac{1}{m^2} \right)$$

$$\leq C'h^{2j-3}. \tag{5.117}$$

Now, we have

$$\|g - \tilde{g}\|_{L^2(\mathbf{R}^2)} \leq \|g - g_h\|_{L^2(\mathbf{R}^2)} + \|g_h - \tilde{g}\|_{L^2(\mathbf{R}^2)}$$

$$\leq \frac{1}{2\pi} \|\hat{g} - \hat{g}_h\|_{L^2(\mathbf{R}^2)} + \|g_h - \tilde{g}\|_{L^2(\mathbf{R}^2)}.$$

Hence, using (5.115), (5.117), we get

$$\|g - \tilde{g}\|_{L^2(\mathbf{R}^2)} \le C(h^{2j} + h^{j-3/2})$$
$$\le C'h^{j-3/2}.$$

This completes the proof of Theorem 5.7.

6 Regularization of some inverse problems in potential theory

In this chapter, we shall discuss the following topics
- Cauchy's problem for the Laplace equation ,
- Determination of surface temperature from borehole measurements (the steady case).

Specifically, we will treat three problems numbered 1 to 3, the first two being formulated in Section 6.2, the third in Section 6.3.

All the problems considered here are ill-posed. They will be regularized by various methods: Tikhonov's method, the method of truncated series expansion and the method of truncated integration. As we shall see, regularization by truncated integration for the problems under consideration is equivalent to regularization by Sinc series. For both truncated expansion and truncated integration, we have obtained explicit error estimates for regularized solutions. As general references for inverse problems in potential theory and partial differential equations let us quote [A], [Is1] and [Is2]. On Cauchy's problem for the Laplace equation there exists a large literature on which we do not attempt to give a survey. Let us only quote the recent paper [CHWY] in which it is treated numerically.

The remainder of the chapter consists of three sections. Section 6.1 deals with the analyticity (in each variable) of harmonic functions. Section 6.2 is devoted to regularizations of Cauchy's problem for the Laplace equation. The final Section 6.3 is devoted to the determination of surface temperature from borehole measurements (the steady case). For simplicity in the presentation, we restrict ourselves to the two-dimensional case.

6.1 Analyticity of harmonic functions

We shall make use of the analyticity (in each variable) of harmonic functions. We have

Theorem 6.1. *Let u be a harmonic function on a domain Ω. Then u is analytic in each variable.*

Proof. Since analyticity is a local property, it is sufficient to prove that each (x_0, y_0) in Ω has a neighborhood $I_1 \times I_2 \subset \Omega$ such that u is analytic in each variable x in I_1 and each variable y in I_2. In fact, let R be a rectangle centered at (x_0, y_0) such that $\bar{R} \subset \Omega$. Let

$$\Gamma(x, y; \xi, \eta) = -\frac{1}{2\pi} \ln \sqrt{(x - \xi)^2 + (y - \eta)^2}.$$

Then for $(\xi, \eta) \neq (x, y)$, $\Gamma(x, y; \xi, \eta)$ is harmonic in (ξ, η). Hence for $(\xi, \eta) \neq (x, y)$, the following identity holds

$$\nabla.(u\nabla\Gamma - \Gamma\nabla u) = 0$$

where the operations of divergence and gradient are with respect to (ξ, η). Applying the divergence theorem on the domain $R\backslash\overline{B}_\epsilon(x, y)$ where $\overline{B}_\epsilon(x, y)$ is the closed ball centered at (x, y) and of radius $\epsilon > 0$ sufficiently small, then letting $\epsilon \to 0$, we have after some rearrangements

$$u(x, y) = -\frac{1}{2\pi} \int_{\partial R} \ln\sqrt{(x - \xi)^2 + (y - \eta)^2}\frac{\partial u}{\partial n}(\xi, \eta)ds$$

$$-\frac{1}{2\pi} \int_{\partial R} \frac{(x - \xi)n_1 + (y - \eta)n_2}{(x - \xi)^2 + (y - \eta)^2}u(\xi, \eta)ds, \quad (x, y) \in R, \tag{6.1}$$

where $ds = ds(\xi, \eta)$ is the arc length differential and $n_1 = n_1(\xi, \eta)$, $n_2 = n_2(\xi, \eta)$ are the x and y-components of the unit outer normal to ∂R.

From (6.1) it is immediately seen that in R, $u(x, y)$ is analytic in x for y fixed and analytic in y for x fixed. This completes the proof of Theorem 6.1.

Remark 6.1. This theorem can also be proved using the fact that a real harmonic function $u(x, y)$ is the real part of an analytic function $w(z)$ with $z = x + iy$. However, we avoid methods of complex analysis, as they essentially are restricted to potential theory in $\mathbf{R}^2$.

Using the analyticity of harmonic functions proved in Theorem 6.1 above, we can formulate Problem 1 and Problem 2 as moment problems. For both problems, we assume that the boundary contains a segment L_0 parallel to the x-axis on which $\eta = k > 0$ (say). We consider a subdomain Ω_1 of Ω (where Ω is a bounded domain in Problem 1 and is an infinite strip in Problem 2) such that $L_0 \subset \partial\Omega_1 \cap \partial\Omega$. Let u be a harmonic function on Ω and let

$$g(x, y; \xi, \eta) = -\frac{1}{2\pi} \ln\sqrt{(x - \xi)^2 + (y - \eta)^2}$$

$$-\frac{1}{2\pi} \ln\sqrt{(x - \xi)^2 + (y + \eta - 2k)^2}. \tag{6.2}$$

Then for $(\xi, \eta) \neq (x, y)$, we have

$$\nabla.(u\nabla g - g\nabla u) = 0 \tag{6.3}$$

where the divergence and gradient operators are with respect to (ξ, η). Let $(x, y) \in \Omega_1$. Integrating (6.3) over $\Omega_1 \backslash \overline{B}_\epsilon(x, y)$ where $\overline{B}_\epsilon(x, y)$ is the closed ball of sufficiently small radius $\epsilon > 0$ centered at (x, y), using the divergence theorem and letting $\epsilon \to 0$, we have after some rearrangements

$$u(x, y) = \int_{L_0} g(x, y; \xi, \eta)v(\xi, \eta)ds$$

$$+ \int_{\partial\Omega_1\backslash L_0} g(x, y; \xi, \eta)v(\xi, \eta)ds$$

$$- \int_{\partial\Omega_1\backslash L_0} \frac{\partial g}{\partial n}(x, y; \xi, \eta)u(\xi, \eta)ds \tag{6.4}$$

where $v = \partial u/\partial n$ and

$$\frac{\partial g}{\partial n}(x, y; \xi, \eta) = n_1(\xi, \eta)\frac{\partial g}{\partial \xi}(x, y; \xi, \eta) + n_2(\xi, \eta)\frac{\partial g}{\partial \eta}(x, y; \xi, \eta)$$

$$= \frac{1}{2\pi}\frac{(x - \xi)n_1(\xi, \eta) + (y - \eta)n_2(\xi, \eta)}{(x - \xi)^2 + (y - \eta)^2}$$

$$+ \frac{1}{2\pi}\frac{(x - \xi)n_1(\xi, \eta) - (y + \eta - 2k)n_2(\xi, \eta)}{(x - \xi)^2 + (y + \eta - 2k)^2}.$$

From (6.4), it is readily seen that if $v = 0$ on L_0, then u is analytic in x for x in L_0. We have thus proved the following

Proposition 6.2. *Let u be a harmonic function in a plane domain Ω such that $\partial\Omega$ contains a segment L_0 parallel to the x-axis. If $\partial u/\partial y = 0$ on L_0 then u is analytic in x for $x \in L_0$.*

Remark 6.2. This proposition can also be proved using results from complex analysis. It is a consequence of the Schwarz reflection principle.

6.2 Cauchy's problem for the Laplace equation

We shall give below important examples in Medicine and in Geophysics of Cauchy's problem for the Laplace equation.

A problem of great interest in Electrocardiology is the computation of the electric potential on a closed surface near and surrounding the heart, given the potential on a part of a body surface and the geometry of the heart and of the thorax. This inverse problem leads to a Cauchy problem for the Laplace equation in the case of constant conductivity.

A mathematical model of the electric field related to the bioelectric activity of the heart is given by Maxwell's equations. It can be shown from the relative values of the coefficients that the time derivatives in the equation can be neglected as a first approximation. Accordingly, the electric field $\mathbf{E}$ and the current density $\mathbf{J}$ satisfy the equations

$$\nabla \times \mathbf{E} = 0, \quad \nabla.\mathbf{J} = 0$$

outside the cardiac region. Now the physiological tissue can be considered a linear resistive medium and hence we can take $\mathbf{J} = \sigma\mathbf{E}$. Since $\mathbf{E}$ is irrotational, it admits a scalar potential V. Therefore with the conductivity σ assumed to be a constant, V satisfies

$$\Delta V = 0 \quad \text{outside the heart.}$$

Thus we arrive at the following Cauchy problem for V in the domain Ω bounded by an outer surface L and an inner surface l :

$$\Delta V = 0 \quad in\ \Omega,$$
$$\frac{\partial V}{\partial n} = 0 \quad on\ L,$$
$$V = f \quad on\ L_0 \subset L$$

where L_0 is an open subset of L. Note that we have taken $\partial V/\partial n = 0$ on the outer boundary of the body since the surrounding medium is air which is electrically nonconducting.

Our second example of a Cauchy problem arises in Gravimetry, the field of Geophysics dealing with the gravity fields in the Earth. Assuming a Flat Earth model, we represent an upper layer of the Earth by the strip

$$\{(x,y): \quad 0 < y < \phi(x), \quad -\infty < x < \infty\},$$

the line $y = \phi(x)$ corresponding to the Earth's surface. The problem is to determine the gravity potential and the gravity field in the strip from gravity data, i.e., the gravity potential and the gravity field measured on all or part of L, the line $y = \phi(x)$, $-\infty < x < \infty$.

We shall consider successively the following problems

Problem 1. Regularization of a Cauchy problem for the Laplace equation in a bounded domain of $\mathbf{R}^2$,

Problem 2. Regularization of a Cauchy problem for the Laplace equation in an irregular strip of $\mathbf{R}^2$.

Before considering each problem individually, we first point out that a general method for dealing with Cauchy's problem for the Laplace equation is the method of quasi-reversibility, introduced by Lattès-Lions (see [LL1] and [LL2]). For the problems under study, exploiting the special geometry of the domains and basing ourselves on the fact that experimental measurements give only finite sets of data, we shall give a simplified treatment, using the moment approach or the truncated integration or truncated expansion approach. Accordingly, both Problem 1 and Problem 2 are formulated as moment problems.

Consider first Problem 1. For this problem we take a plane domain Ω bounded by two C^1-Jordan curves, L and l, such that l is interior to L. It is assumed throughout this discussion that L contain a segment L_0 parallel to the x-axis. Let u be a harmonic function on Ω such that $u \in C^1(\Omega \cup L_0)$. The problem is to determine $\partial u/\partial n$, the normal derivative of u, on the interior boundary l from the values of $\partial u/\partial n$ on L and values of u on an appropriate subset of L_0. This problem will be formulated as a moment problem. In fact we have

Theorem 6.3. *Let $N(x, y; \xi, \eta)$ be the Neumann function for the Laplace equation in the domain Ω described above. Let u be a harmonic function on Ω such that $u \in C(\overline{\Omega})$ and is piecewise C^1 on $\partial\Omega$. Then u admits the representation*

$$u(x,y) = a + \int_{L_0} N(x,y;\xi,\eta)v(\xi,\eta)ds(\xi,\eta) +$$

$$+ \int_{L\backslash L_0} N(x,y;\xi,\eta)v(\xi,\eta)ds(\xi,\eta) +$$

$$+ \int_l N(x,y;\xi,\eta)v(\xi,\eta)ds(\xi,\eta) \tag{6.5}$$

where $v = \partial u/\partial n$, a is a constant to be determined and ds is the arc length differential along $\partial\Omega$.

Let v be given on L and u given on a bounded sequence of points $(x_n, y_n) \in L_0$ such that $y_n = k$ $\forall n$ and $x_i \neq x_j$ for $i \neq j$. Then the moment problem for a and v

$$\int_l N(x_n, y_n; \xi, \eta) v(\xi, \eta) ds(\xi, \eta) =$$

$$= u(x_n, y_n) - a - \int_{L_0} N(x_n, y_n; \xi, \eta) v(\xi, \eta) ds(\xi, \eta)$$

$$- \int_{L \setminus L_0} N(x_n, y_n; \xi, \eta) v(\xi, \eta) ds(\xi, \eta), \qquad n = 1, 2, \ldots, \qquad (6.6)$$

admits at most one solution.

Proof. Let u_1, u_2 be two harmonic functions on Ω satisfying (6.6). Then, by Proposition 6.2, $u_1 - u_2$ is analytic on L_0 since $\partial(u_1 - u_2)/\partial n = 0$ on L_0. By the identity theorem for analytic functions , $u_1 - u_2 = 0$ on L_0. By the uniqueness of Cauchy's problem for harmonic functions, $u_1 - u_2 = 0$ on Ω. It follows that

$$\frac{\partial u_1}{\partial n} = \frac{\partial u_2}{\partial n} \quad \text{on } l$$

i.e. $v_1 = v_2$ on l. This proves uniqueness for the moment problem and thus completes the proof of Theorem 6.3.

Regularization of Problem 1.
Let

$$L_0 = \{(x, k) : \ 0 < x < 1\}, \quad (x_n, y_n) = (x_n, k) \in L_0, \quad x_j \neq x_m \text{ for } j \neq m.$$

We assume that there are two C^1 functions of the parameter $t \in [0, 1]$ such that l has the representation

$$l(t) = (\xi(t), \eta(t)) \quad \forall t \in [0, 1], \quad l(0) = l(1).$$

Using these functions, we can rewrite (6.6) as

$$\int_0^1 N(x_n, k; \xi(t), \eta(t)) \sqrt{|\xi'(t)|^2 + |\eta'(t)|^2} v(\xi(t), \eta(t)) dt =$$

$$= u(x_n, k) - a - \int_L N(x_n, k; \xi, \eta) v_L(\xi, \eta) ds(\xi, \eta),$$

$$n = 1, 2, \ldots, \qquad (6.7)$$

where v_L is given.

We denote by H the Hilbert space

$$H = \{(a, \psi) : \ a \in \mathbf{R}, \ \psi \in H^1(0, 1), \ \psi(0) = \psi(1)\}$$

with the norm

$$\|(a, \psi)\|_H^2 = a^2 + \|\psi\|_{H^1(0,1)}^2.$$

We define $A : H \to l^2$ by

$$A(a, \psi) = \left(\frac{a}{n} + \frac{1}{n} \int_0^1 N(x_n, k; \xi(t), \eta(t)) \sqrt{|\xi'(t)|^2 + |\eta'(t)|^2} \psi(t) dt \right)_{n \geq 1}.$$

Using these notations, we can rewrite (6.7) (or (6.8)) as

$$A(a, z) = \mu, \qquad (a, z) \in H, \tag{6.8}$$

where $\mu = (\mu_n)_{n \geq 1}$ and

$$\mu_n = \frac{1}{n} \left(u(x_n, k) - \int_L N(x_n, k; \xi, \eta) v(\xi, \eta) ds(\xi, \eta) \right),$$

$$z(t) = v(\xi(t), \eta(t)) \quad \left(= \frac{\partial u}{\partial n} \right).$$

We shall use the method described in Chapter 2, Subsection 2.2.3 to regularize (6.8). For $\beta > 0$, we consider the problem of finding $w^\beta = w^\beta(\tilde{\mu}) \in H$ such that

$$\beta(w^\beta, w)_H + (Aw^\beta, Aw)_{l^2} = (\tilde{\mu}, Aw)_{l^2}, \qquad \forall w \in H, \tag{6.9}$$

where $(.,.)_H$, $(.,.)_{l^2}$ are the inner products of H, l^2 respectively. The following theorem follows directly from the results in Subsection 2.2.3, Chapter 2.

Theorem 6.4. *Let $\epsilon > 0$. Let $w_0 = (a, z)$ be the exact solution of (6.8) corresponding to $\mu^0 \in l^2$ in the right hand side.*

Then there exists a continuous increasing function

$$\varphi : [0, \infty) \longrightarrow [0, \infty)$$

with

$$\lim_{t \downarrow 0} \varphi(t) = \varphi(0) = 0,$$

such that, for all $\mu \in l^2$, satisfying

$$\|\mu - \mu^0\|_{l^2} < \epsilon,$$

we have the estimate

$$\left(|a_\epsilon - a|^2 + \|z_\epsilon - z\|_{L^2(0,1)}^2 \right)^{1/2} \leq \varphi(\epsilon^{1/2})$$

where $(a_\epsilon, z_\epsilon) = w^\epsilon$ is the solution of (6.9) with $\beta = \epsilon$.

We now turn to Problem 2, i.e., the Cauchy problem for the Laplace equation in the plane domain

$$\Omega = \{(x, y) : 0 < y < \phi(x), -\infty < x < \infty\}$$

the Cauchy data being prescribed on the curve $L : y = \phi(x)$. The problem is to find $v(x) = u(x, 0)$ on the line $l : y = 0$. The curve L is assumed to be smooth and eventually straight, i.e., with constants c_+, c_- we have $\phi(x) = c_+$ for $x > 0$

sufficiently large, $\phi(x) = c_-$ for $x < 0$ and $-x$ sufficiently large,. The problem could be formulated as a moment problem in the same manner as for Problem 1, and the moment problem could also be regularized using the Tikhonov method. However, we no longer have at our disposal a Sobolev compact embedding theorem and as a consequence the arguments we have used to estimate the convergence of regularized solutions do not apply here. We shall not pursue this matter any further, but, instead, will follow the integral equation approach after having patched up the given Cauchy data taken at equidistant points on **R** into a Sinc series.

We can now give a precise formulation of the problem. We first set some notations. For $y = \phi(x)$ let

$$u_x(x,y) = f(x), \ u_y(x,y) = g(x), \ u(x,y) = w(x). \tag{6.10}$$

We remark at once that these are the *exact values* of u_x, u_y and u on L, which need not be known. All that we may know are the *measured values* taken at certain points of L. Following are our standing assumptions

A1. $f(x)$ and $g(x)$ are in $C(\mathbf{R})$, $w(x)$ is in $C^2(\mathbf{R})$,
A2. $f(x), g(x), w(x)$ tend to zero at least as $1/|x|$ for $|x| \to \infty$ and

$$(1+x^2)^{1/2}w', \ (1+x^2)^{1/2}w'' \text{ are in } L^2(\mathbf{R}),$$

A3. $(1+x^2)^{1/2}v$ is in $L^2(\mathbf{R})$, $v(x) = u(x,0)$.

From (6.10), we get a Fredholm equation in $v(x)$. In fact, put

$$G(x,y;\xi,\eta) = \Gamma(x,y;\xi,\eta) - \Gamma(x,y;\xi,-\eta)$$

where, we recall

$$\Gamma(x,y;\xi,\eta) = -\frac{1}{2\pi} \ln \sqrt{(x-\xi)^2 + (y-\eta)^2}.$$

Using G, we can derive an integral equation for the function $v(x) = u(x,0)$. Indeed, let $\Omega_\epsilon = \Omega \setminus \overline{B}_\epsilon(x,y)$ ($\overline{B}_\epsilon(x,y)=$ closed ball of sufficiently small radius ϵ centered at (x,y)). Integrating the identity

$$div(G\nabla u - u\nabla G) = 0 \quad \forall (\xi,\eta) \in \Omega_\epsilon$$

on Ω_ϵ and letting $\epsilon \downarrow 0$ give in view of (A1)-(A3)

$$\frac{1}{\pi} \int_{-\infty}^{\infty} \frac{yv(\xi)}{(x-\xi)^2 + y^2} d\xi = -u(x,y)$$
$$- \int_{-\infty}^{\infty} G(x,y;\xi,\phi(\xi))(g(\xi) - f(\xi)\phi'(\xi))d\xi$$
$$+ \int_{-\infty}^{\infty} \frac{\partial G}{\partial n}(x,y;\xi,\phi(\xi))w(\xi)ds(\xi,\phi(\xi))$$
$$\text{for } 0 < y < \phi(x), \ -\infty < x < \infty, \tag{6.11}$$

where

$$\frac{\partial G}{\partial n}(.;\xi,\phi(\xi)) = -\frac{\partial G}{\partial \xi}(.;\xi,\phi(\xi))\frac{\phi'(\xi)}{\sqrt{1+|\phi(\xi)|^2}}$$

$$+\frac{\partial G}{\partial \eta}(.;\xi,\phi(\xi))\frac{1}{\sqrt{1+|\phi(\xi)|^2}}.$$

In order to have an integral equation in $v(x)$, we let $y \uparrow \phi(x)$. We shall make use of the jump relation (see, e.g., [Co3], Chap.5)

$$\lim_{y\uparrow\phi(x)} \int_{-\infty}^{\infty} \frac{\partial G}{\partial n}(x,y;\xi,\phi(\xi))w(\xi)ds(\xi,\phi(\xi)) =$$

$$= -\frac{1}{2} w(x) + \int_{-\infty}^{\infty} \frac{\partial G}{\partial n}(x,\phi(x);\xi,\phi(\xi))w(\xi)ds(\xi,\phi(\xi)). \qquad (6.12)$$

Letting $y \uparrow \phi(x)$ in (6.11) gives in view of (6.12)

$$\frac{1}{\pi} \int_{-\infty}^{\infty} \frac{\phi(x)v(\xi)}{(x-\xi)^2+\phi^2(x)}d\xi = -\frac{3}{2}w(x)$$

$$-\int_{-\infty}^{\infty} G(x,\phi(x);\xi,\phi(\xi))(g(\xi)-f(\xi)\phi'(\xi))d\xi$$

$$+\int_{-\infty}^{\infty} \frac{\partial G}{\partial n}(x,\phi(x);\xi,\phi(\xi))w(\xi)ds(\xi,\phi(\xi)),$$

hence the integral equation

$$\frac{1}{\pi} \int_{-\infty}^{\infty} \frac{\phi(x)v(\xi)}{(x-\xi)^2+\phi^2(x)}d\xi = -\frac{3}{2}w(x)$$

$$-\int_{-\infty}^{\infty} G(x,\phi(x);\xi,\phi(\xi))(g(\xi)-f(\xi)\phi'(\xi))d\xi$$

$$-\int_{-\infty}^{\infty} \frac{\partial G}{\partial \xi}(x,\phi(x);\xi,\phi(\xi))\phi'(\xi)w(\xi)d\xi$$

$$+\int_{-\infty}^{\infty} \frac{\partial G}{\partial \eta}(x,\phi(x);\xi,\phi(\xi))w(\xi)d\xi, \qquad (6.13)$$

which is an integral equation of the first kind.

Using the method of Tikhonov, one could regularize this equation. However it is usually difficult to derive an error estimate between a regularized solution and the exact solution. Hence, we shall transform (6.13) into a convolution equation in $v(x)$, for which error estimates are more easily derived. Note that the function in the left hand side of (6.11) is harmonic in the half plane $y > 0$. Denote by H that function, i.e.

$$H(x,y) = \frac{1}{\pi} \int_{-\infty}^{\infty} \frac{yv(\xi)}{(x-\xi)^2+y^2}d\xi.$$

For $0 < y < \phi(x)$, $H(x,y)$ is given by (6.11). We shall calculate $H(x,y)$ for $y > \phi(x)$ in terms of w, f, g. Since the limit of $H(x,y)$ as $(x,y) \to \infty$ is zero, the values of $H(x,y)$ for $y > \phi(x)$ are determined uniquely from $H(x,\phi(x))$, $\frac{\partial H}{\partial n}(x,\phi(x))$.

In (6.13), $H(x, \phi(x))$ is given in terms of w, f, g. On the other hand, $H(x, y)$ has continuous derivatives everywhere in the domain $y > 0$ and in particular for $y \geq \phi(x)$. Hence

$$\frac{\partial H}{\partial n}(x, \phi(x)) = \lim_{y \uparrow \phi(x)} \frac{\partial H}{\partial n}(x, y) = \frac{1}{\pi} \lim_{y \uparrow \phi(x)} \frac{\partial}{\partial n} \int_{-\infty}^{\infty} \frac{yv(\xi)}{(x - \xi)^2 + y^2} d\xi$$

where, in this case, we define

$$\frac{\partial \varphi}{\partial n}(x, y) = \frac{\partial \varphi}{\partial x}(x, y) \frac{\phi'(x)}{\sqrt{1 + |\phi'(x)|^2}} - \frac{\partial \varphi}{\partial y}(x, y) \frac{1}{\sqrt{1 + |\phi'(x)|^2}}.$$

Thus, using (6.11) it is possible to calculate $\frac{\partial H}{\partial n}(x, \phi(x))$ in terms of w, f, g. As mentioned above, $H(x, y)$ is defined completely in terms of w, f, g for all $y > \phi(x)$. In particular for $y = k > \sup_{x \in \mathbf{R}} \phi(x)$, we have

$$\frac{1}{\pi} \int_{-\infty}^{\infty} \frac{kv(\xi)}{(x - \xi)^2 + k^2} d\xi = F(x; f, g, w)$$

where F is the function given by (6.17). This is the desired convolution equation in $v(x)$. We will regularize this convolution equation. But first we derive an explicit formula for $F(x; f, g, w)$ (i.e. (6.17)). The details are somewhat tedious, so the reader can simply skip them at a first reading and jump to the formula (6.17).

For the derivation of (6.17), integrate Green's Identity

$$div\,(\Gamma \nabla H - H \nabla \Gamma) = 0$$

in the domain $\Omega_R \setminus \overline{B}_\epsilon(x, y)$, where

$$\Omega_R = \{(\xi, \eta) : |\xi| < R,\ \phi(\xi) < \eta < R\}.$$

Letting $R \to \infty$, $\epsilon \to 0$, we get

$$H(x, y) = \int_{-\infty}^{\infty} \Gamma(x, y; \xi; \phi(\xi)) \frac{\partial H}{\partial n}(\xi, \phi(\xi)) ds(\xi, \phi(\xi))$$
$$- \int_{-\infty}^{\infty} \frac{\partial \Gamma}{\partial n}(x, y; \xi; \phi(\xi)) H(\xi, \phi(\xi)) ds(\xi, \phi(\xi))$$

where $n = n(\xi, \phi(\xi)) = (1 + |\phi'(\xi)|^2)^{-1/2}(\phi'(\xi), -1)$ is the unit outer normal to the boundary of the domain

$$\{(\xi, \eta) : \phi(\xi) < \eta,\ -\infty < \xi < \infty\}$$

and

$$ds(\xi, \phi(\xi)) = (1 + |\phi'(\xi)|^2)^{1/2} d\xi.$$

Rewriting (6.13) somewhat, one has

$$H(x, \phi(x)) = -\frac{3}{2} w(x)$$
$$- \int_{-\infty}^{\infty} G(x, \phi(x); \xi, \phi(\xi))(g(\xi) - f(\xi)\phi'(\xi))d\xi$$
$$- \int_{-\infty}^{\infty} \frac{\partial G}{\partial \xi}(x, \phi(x); \xi, \phi(\xi))\phi'(\xi)w(\xi)d\xi$$
$$+ \int_{-\infty}^{\infty} \frac{\partial G}{\partial \eta}(x, \phi(x); \xi, \phi(\xi))w(\xi)d\xi. \tag{6.14}$$

Taking the derivatives of the terms in (6.11) in the direction of the vector $n(x, \phi(x))$, letting $y \uparrow \phi(x)$ in the result thus obtained and using the jump relation (6.12), we get after some rearrangements

$$\frac{\partial H}{\partial n}(x, \phi(x)) = \lim_{y \uparrow \phi(x)} \frac{\partial}{\partial n} \int_{-\infty}^{\infty} \frac{yv(\xi)}{(x - \xi)^2 + y^2} d\xi$$
$$= \frac{3}{2}(g(x) - f(x)\phi'(x))$$
$$- \frac{1}{\alpha(x)} \int_{-\infty}^{\infty} G_1(x, \phi(x); \xi, \phi(\xi))(g(\xi) - f(\xi)\phi'(\xi))d\xi$$
$$+ \frac{1}{\alpha(x)} \int_{-\infty}^{\infty} G_2(x, \phi(x); \xi, \phi(\xi))w'(\xi)d\xi$$
$$+ \frac{1}{4\pi\alpha(x)} \int_{-\infty}^{\infty} \ln \sqrt{(x - \xi)^2 + (\phi(x) - \phi(\xi))^2}\, w''(\xi)d\xi \tag{6.15}$$

where

$$G_1(x, y; \xi, \eta) = G_x(x, y; \xi, \eta)\phi'(x) - G_y(x, y; \xi, \eta),$$
$$G_2(x, y; \xi, \eta) = \frac{1}{2\pi} \frac{(y + \eta)\phi'(x) + (x - \xi)}{(x - \xi)^2 + (y + \eta)^2} +$$
$$+ \frac{1}{2\pi} \frac{(y - \eta)(\phi'(x) - \phi'(\xi))}{(x - \xi)^2 + (y - \eta)^2},$$
$$\alpha(x) = \sqrt{1 + |\phi'(x)|^2}.$$

Now, letting $y = k > \sup_{x \in \mathbf{R}} \phi(x)$ in (6.11), one gets the equation

$$\frac{1}{\pi} \int_{-\infty}^{\infty} \frac{kv(\xi)}{(x - \xi)^2 + k^2} d\xi = F(x; f, g, w) \tag{6.16}$$

where

$$F(x; f, g, w) = \int_{-\infty}^{\infty} \Gamma(x, k; \xi, \phi(\xi)) \frac{\partial H}{\partial n}(\xi, \phi(\xi)) ds(\xi, \phi(\xi))$$
$$- \int_{-\infty}^{\infty} \frac{\partial \Gamma}{\partial n}(x, k; \xi, \phi(\xi)) H(\xi, \phi(\xi)) ds(\xi, \phi(\xi)) \tag{6.17}$$

and $H(x, \phi(x))$, $\frac{\partial H}{\partial n}(x, \phi(x))$ are given in terms of w, f, g by (6.14),(6.15).

The convolution equation (6.16) determines v uniquely if we have the exact values of $F(x; f, g, w)$. However, the functions f, g, w are not known exactly. Hence, $F(x; f, g, w)$ is not known exactly. As mentioned earlier we only know the measured values of f, g, w at certain points of L and furthermore these are affected with noise. We shall assume that these values are given on points of L having equidistant x-coordinates. Hence, using Sinc expansion as in Chapter 5, under some smoothness assumptions on f, g, w, we can patch up these values into functions approximating f, g, w in the $L^2(\mathbf{R})$-sense. Accordingly, the function $F(x; f, g, w)$ will be approximated by a function constructed from the measured values.

To be specific, for $h > 0$, we shall assume that (ξ_n), (μ_n), (ν_n) are sequences in $l^2(\mathbf{Z})$ such that the quantity

$$\sum_{n\in\mathbf{Z}} \left\{ |f(nh) - \mu_n|^2 + |g(nh) - \nu_n|^2 + |w(nh) - \xi_n|^2 \right\}$$

is small. Using the result in Subsection 5.3, Chapter 5, we can approximate f, g, w by

$$f_h(x) = \sum_{n\in\mathbf{Z}} \mu_n S(n, h)(x),$$

$$g_h(x) = \sum_{n\in\mathbf{Z}} \nu_n S(n, h)(x),$$

$$w_h(x) = \sum_{n\in\mathbf{Z}} \xi_n S(n, h)(x).$$

For the proof of Theorem 6.5, a main result of this chapter (whose statement will be given later), we rely on the following

Lemma 6.1. *Let $(\mu_n)_{n\in\mathbf{Z}}$, $(\nu_n)_{n\in\mathbf{Z}}$, $(\xi_n)_{n\in\mathbf{Z}}$ be in $l^2(\mathbf{Z})$ and let $h > 0$. Suppose that*

a) the functions f, g, w are in the set

$$W = \{\psi \in L^2(\mathbf{R}) : \ (1 + \omega^2)\hat{\psi} \in L^2(\mathbf{R}) \cap L^\infty(\mathbf{R}), \ \ (1 + \omega^2)\hat{\psi}' \in L^2(\mathbf{R})\}$$

where $\hat{\psi}$ is the Fourier transform of ψ,

$$\hat{\psi}(\omega) = \int_{-\infty}^{\infty} \psi(x)e^{ix\omega}\,dx,$$

b) f, g, w, w, w', w'' are in the set

$$\{\psi \in L^2(\mathbf{R}) : \ (1 + x^2)\psi \in L^\infty(\mathbf{R})\}$$

and $w''' \in L^2(\mathbf{R})$,

c) $(1 + x^2)v \in L^2(\mathbf{R})$.

Assume further that

$$\sum_{n\in\mathbf{Z}} \left\{ |f(nh) - \mu_n|^2 + |g(nh) - \nu_n|^2 + |w(nh) - \xi_n|^2 \right\} \le Ch \qquad (6.18)$$

for a constant C independent from h.

Then there is a function $F_h \equiv F_h(f_h, g_h, w_h)$ such that

$$\|F - F_h\|_{L^2(\mathbf{R})} \leq C_0 \sqrt[50]{h}$$

where $F \equiv F(x; f, g, w)$ is as in (6.17) and C_0 is a (computable) constant independent from h.

Sketch of the proof of the lemma.
From Theorem 5.5, we infer that

$$\|f - f_h\|_{L^2(\mathbf{R})} + \|g - g_h\|_{L^2(\mathbf{R})} + \|w - w_h\|_{L^2(\mathbf{R})} \leq Ch. \qquad (6.19)$$

For the purpose of the proof, we shall approximate various quantities.
Approximation of $H(x, \phi(x)) \equiv H_1(x)$:
Put

$$
\begin{aligned}
H_h(x) = &-\frac{3}{2} w_h(x) \\
&- \int_{-2\sqrt[5]{h^{-4}}}^{2\sqrt[5]{h^{-4}}} G(x, \phi(x); \xi, \phi(\xi))(g_h - f_h(\xi)\phi'(\xi))d\xi \\
&- \int_{-2\sqrt[5]{h^{-4}}}^{2\sqrt[5]{h^{-4}}} \frac{\partial G}{\partial \xi}(x, \phi(x); \xi, \phi(\xi))\phi'(\xi)w(\xi)d\xi \\
&+ \int_{-2\sqrt[5]{h^{-4}}}^{2\sqrt[5]{h^{-4}}} \frac{\partial G}{\partial \eta}(x, \phi(x); \xi, \phi(\xi))w(\xi)d\xi \\
&\qquad \text{for } |x| < \sqrt[10]{h^{-4}}, \\
H_h(x) = 0 \qquad &\forall |x| \geq \sqrt[10]{h^{-4}}.
\end{aligned}
$$

By direct computations, one has

$$\|H_1 - H_h\|_{L^2(\mathbf{R})} \leq C' \sqrt[5]{h}. \qquad (6.20)$$

Approximation of $\frac{\partial H}{\partial n}(x, \phi(x)) \equiv H_2(x)$:
Put

$$
\begin{aligned}
K_h(x) = &\frac{3}{2}(g_h(x) - f_h(x)\phi'(x)) \\
&- \frac{1}{\alpha(x)} \int_{-2\sqrt[5]{h^{-4}}}^{2\sqrt[5]{h^{-4}}} G_2(x, \phi(x); \xi, \phi(\xi))(g_h - f_h(\xi)\phi'(\xi))d\xi \\
&+ \frac{1}{\alpha(x)} \int_{-2\sqrt[5]{h^{-4}}}^{2\sqrt[5]{h^{-4}}} G_3(x, \phi(x); \xi, \phi(\xi))w_{1h}(\xi))d\xi \\
&\frac{1}{4\pi\alpha(x)} \int_{-2\sqrt[5]{h^{-4}}}^{2\sqrt[5]{h^{-4}}} \ln\left((x - \xi)^2 + (\phi(x) - \phi(\xi))^2\right)^{1/2} w_{2h}(\xi))d\xi \\
&\qquad \text{for } |x| < \sqrt[10]{h^{-4}}, \\
K_h(x) = 0 \qquad &\text{for } |x| \geq \sqrt[10]{h^{-4}}
\end{aligned}
$$

where

$$w_{1h} = f_h + \phi' g_h$$

$$w_{2h}(x) = \frac{w_{1h}(x + \sqrt{h}) - w_{1h}(x))}{\sqrt{h}}.$$

By direct computations one has

$$\|K_h - H_2\|_{L^2(\mathbf{R})} \leq C' \sqrt[10]{h}. \tag{6.21}$$

Approximation of $F \equiv F(x; f, g, w)$:
Put

$$F_h(x) = \int_{-2\sqrt[25]{h^{-4}}}^{2\sqrt[25]{h^{-4}}} \Gamma(x, k; \xi; \phi(\xi)) K_h(\xi) ds(\xi, \phi(\xi))$$

$$- \int_{-2\sqrt[25]{h^{-4}}}^{2\sqrt[25]{h^{-4}}} \frac{\partial \Gamma}{\partial n}(x, k; \xi; \phi(\xi)) H_h(\xi) ds(\xi, \phi(\xi))$$

$$\text{for } |x| < \sqrt[25]{h^{-1}},$$

$$F_h(x) = 0 \quad \text{for } |x| \geq \sqrt[25]{h^{-1}}.$$

Then, in view of (6.20), (6.21) we get after some computations

$$\|F - F_h\|_{L^2(\mathbf{R})} \leq C_0 \sqrt[50]{h}. \tag{6.22}$$

This completes the proof of Lemma 6.1.

Using Lemma 6.1 we shall construct a regularized solution of (6.16) admitting a Sinc series representation. We make the following assumption on the exact solution v of (6.16) corresponding to $F(x; f, g, w)$ in the right hand side

$$\|t\hat{v}(t)\|_{L^2(\mathbf{R})}^2 \leq E^2 \tag{6.23}$$

where $\hat{v}$ is the Fourier transform of v. Condition (6.23) means that v is in $H^1(\mathbf{R})$, the $H^1(\mathbf{R})$-norm of which is majorized by E. Let the kernel of Eq. (6.16) (i.e. the Cauchy kernel) be denoted by $K(x)$,

$$K(x) = \frac{k}{\pi(x^2 + k^2)}. \tag{6.24}$$

The Fourier transform $\hat{K}(t)$ of $K(x)$ is

$$\hat{K}(t) = \int_{-\infty}^{\infty} K(x) e^{ixt} dx = e^{-k|t|}. \tag{6.25}$$

Let $\epsilon = \sqrt[50]{h}$ and consider the function

$$v_\epsilon(x) = \frac{1}{2\pi} \int_{|t| \leq t_{\epsilon_0}} \hat{F}_h(t) e^{-k|t|} e^{-ixt} dt \tag{6.26}$$

where $\epsilon_0 = C_0\epsilon$ with C_0 as in (6.22), and

$$t_{\epsilon_0} = \frac{1}{k}\ln\left(\frac{E}{\epsilon_0}\left(\ln\frac{E}{\epsilon_0}\right)^{-1}\right). \tag{6.27}$$

Note that $\hat{v}_\epsilon(t)$ has compact support, supp $\hat{v}_\epsilon(t) \subset [-t_{\epsilon_0}, t_{\epsilon_0}]$ and thus $v_\epsilon(x)$ can be represented by a Sinc series. We shall take v_ϵ as our regularized solution. The following theorem gives an estimate of the error between v_ϵ and the exact solution of (6.16).

Theorem 6.5. *For $n \in \mathbf{Z}$, let μ_n, ν_n, ξ_n be the measured values of $f(nh), g(nh)$ and $w(nh)$ respectively. Let $\epsilon > 0$ be given and let $h = \epsilon^{50}$. Suppose that (6.18) holds and that the exact solution of (6.16) satisfies (6.23). Then, for $\epsilon \to 0$, there holds the estimate*

$$\|v_\epsilon - v\|_{L^2(\mathbf{R})} \leq \sqrt{\frac{1+k^2}{2\pi}}\frac{1}{\ln\left(\frac{E}{\epsilon_0}\left(\ln\frac{E}{\epsilon_0}\right)^{-1}\right)}$$

where C_0 is the constant in (6.22)

Proof. We have

$$\|\hat{v}_\epsilon - \hat{v}\|_{L^2(\mathbf{R})}^2 = \int_{|t|\leq t_{\epsilon_0}}|\hat{v}_\epsilon - \hat{v}|^2 dt + \int_{|t|>t_{\epsilon_0}}|\hat{v}|^2 dt. \tag{6.28}$$

Now

$$\int_{|t|\leq t_{\epsilon_0}}|\hat{v}_\epsilon - \hat{v}|^2 dt = \int_{|t|\leq t_{\epsilon_0}}|\hat{F} - \hat{F}_h|^2 e^{2kt} dt$$

$$\leq e^{2kt_{\epsilon_0}}\int_{-\infty}^{\infty}|\hat{F} - \hat{F}_h|^2 dt$$

$$= E^2\left(\ln\frac{E}{\epsilon_0}\right)^{-2}. \tag{6.29}$$

On the other hand

$$\int_{|t|>t_{\epsilon_0}}|\hat{v}|^2 dt \leq \frac{1}{t_{\epsilon_0}^2}\int_{|t|>t_{\epsilon_0}}|t^2\hat{v}|^2 dt$$

$$\leq \frac{E^2 k^2}{\ln^2\left(\frac{E}{\epsilon_0}\left(\ln\frac{E}{\epsilon_0}\right)^{-1}\right)}. \tag{6.30}$$

Since for small $\epsilon_0 > 0$

$$\ln\left(\frac{E}{\epsilon_0}\left(\ln\frac{E}{\epsilon_0}\right)^{-1}\right) > \left(\ln\frac{E}{\epsilon_0}\right)^{-1}$$

(6.28)-(6.30) yield

$$\|\hat{v}_{\epsilon} - \hat{v}\|^2_{L^2(\mathbf{R})} \leq \frac{(1+k^2)E^2}{\ln^2\left(\frac{E}{\epsilon_0}\left(\ln\frac{E}{\epsilon_0}\right)^{-1}\right)} \qquad \text{for small } \epsilon_0. \qquad (6.31)$$

By Fourier inversion we shall get the desired estimates. This completes the proof of Theorem 6.5.

6.3 Surface temperature determination from borehole measurements (steady case)

In order to determine the temperature on the surface of the Earth, it is useful to make measurements in its interior rather than on the surface since measurements made on the surface are likely to be affected by noise. On the other hand, to compute the surface temperature from interior observations is an ill-posed problem that needs to be regularized.

Let the Earth be represented by a half plane $-\infty < x < \infty$, $y \geq 0$, assuming a Flat Earth model. Such model has, e.g., been used in [GV] in a seismological problem. We shall consider the following

Problem 3. Determination of surface temperature from borehole measurements (the steady case).

Regularization of Problem 3:
We propose to determine the surface temperature $u(x,0)$ from temperature measurements at $y = 1$. Note that in the steady case, the temperature function $u(x,y)$ satisfies the Laplace equation

$$\Delta u = 0, \qquad (x,y) \in \mathbf{R} \times \mathbf{R}_+. \qquad (6.32)$$

As shown in Sect. 6.1, u is analytic in x for fixed y. Hence, in particular, for $y = 1$, $u(x,1)$ is completely determined by its values on any bounded sequence $(x_n, 1)$, x_n real, $n = 1, 2, \ldots$ and $x_i \neq x_j$ for $i \neq j$. The problem of determining $u(x,0)$ can therefore be formulated as a moment problem as follows.

It can be shown by the Fourier method or the method of Green's function that $u(x,y)$ satisfies the equation

$$u(x,y) = \frac{y}{\pi} \int_{-\infty}^{\infty} \frac{v(\xi)d\xi}{(x-\xi)^2 + y^2}, \qquad y > 0.$$

At $y = 1$, $x = x_n$, we have

$$\frac{1}{\pi} \int_{-\infty}^{\infty} \frac{v(\xi)d\xi}{(x_n - \xi)^2 + 1} = u(x_n, 1), \qquad n = 1, 2, \ldots, \qquad (6.33)$$

where we have set $v(x) = u(x,0)$ and where (x_n) is any bounded real sequence with $x_i \neq x_j$ for $i \neq j$. The moment problem (6.33) has at most one solution in $L^2(\mathbf{R})$. Indeed, if

$$\frac{1}{\pi} \int_{-\infty}^{\infty} \frac{v(\xi)d\xi}{(x_n - \xi)^2 + 1} = 0, \quad n = 1, 2, \ldots,$$

then by the identity theorem for analytic functions

$$\frac{1}{\pi} \int_{-\infty}^{\infty} \frac{v(\xi)d\xi}{(x - \xi)^2 + 1} = 0, \quad \forall x \in \mathbf{R},$$

which implies $v = 0$ in $L^2(\mathbf{R})$. This prove our claim.

If (x_n) is an arbitrary bounded sequence with $x_i \neq x_j$ for $i \neq j$, then we can regularize the moment problem (6.33), e.g., using the Tikhonov method. However deriviny estimates of the error between a regularized solution and the exact solution is usually quite involved. Hence, it is simpler to consider the integral equation

$$\frac{1}{\pi} \int_{-\infty}^{\infty} \frac{v(\xi)d\xi}{(x - \xi)^2 + 1} = u(x, 1), \quad \forall x \in \mathbf{R}, \tag{6.34}$$

and regularize it by Sinc series. The details of the regularization procedure, which are similar to (in fact, simpler than) for Eq.(6.16) considered earlier, are omitted. It is of interest to note that a regularization by Sinc series requires that the right hand side of (6.34) be given (possibly with noise)at equidistant points nh, $n \in \mathbf{Z}$, $h > 0$, of $\mathbf{R}$, thus approximating the original integral equation by a moment problem.

7 Regularization of some inverse problems in heat conduction

This Chapter is concerned with direct applications of the results of previous chapters on moment theory to some inverse problems in Heat Conduction. Many inverse problems in Heat Conduction can be formulated as moment problems, however, we consider here only a few of them, those for which error estimates for regularized solutions can be readily derived using the techniques of the preceding chapters. In the "Notes and remarks" at the end of the chapter, we list some problems and results related to those of the main text.

In this chapter, we shall consider three problems related to the heat equation

$$\Delta u - u_t = 0 \qquad \forall (x, y, t) \in \mathbf{R}^2 \times \mathbf{R}_+ .$$

Problem 1. Determination of the temperature in $\mathbf{R}^2$ at time $t = 0$ from a given temperature at time $t = 1$ (the backward heat equation problem).

Problem 2. Determination of surface temperature from borehole measurements.

Problem 3. Determination of the boundary value $u(x, 0, t)$ assuming known the free boundary and the initial temperature $u_0(x, y)$ (the so-called inverse Stefan problem).

The moment method will be used for Problem 1 and Problem 2 while the method of Sinc expansion will be used for Problem 3.

The remainder of the chapter is divided into three sections corresponding to these three problems.

7.1 The backward heat equation

Consider the heat equation

$$\Delta u - u_t = 0 \qquad \forall (x, y, t) \in \mathbf{R}^2 \times \mathbf{R}_+$$

where we have taken the heat conductivity to be 1 and $u = u(x, y, t)$ denotes the temperature.

We address the problem of determining the initial temperature $v(x, y) = u(x, y, 0)$, given the temperature at $t = 1$. As is well-known, this is an ill-posed problem. Various methods of regularization have been used. One of the better known is the method of quasi-reversibility due to Lattès-Lions [LL1],[LL2] (see also Showalter-Ting [STi] for some variants of the method). In Ang [An1], the problem is formulated as an integral equation of the first kind which is regularized by the Tikhonov method. More recently, the regularization was carried out by a truncated series expansion [AH]. A wide survey on treatment of inverse heat conduction problems with many references is given by Dinh Nho Hào [DNH].

In the present section, we shall consider two methods for regularizing our problem: the method of moments and the method of truncated integration.

We first formulate the problem as an integral equation. Let

$$G(x, y, t; \xi, \eta, \tau) = \frac{1}{4\pi(t - \tau)} \exp \left\{ -\frac{(x - \xi)^2 + (y - \eta)^2}{4(t - \tau)} \right\}. \tag{7.1}$$

Noting that

$$G_{\xi\xi} + G_{\eta\eta} + G_{\tau} = 0$$

we get for $u = u(\xi, \eta, \tau)$

$$\operatorname{div}(u\nabla G - G\nabla u) + (Gu)_{\tau} = 0 \tag{7.2}$$

Integrate the latter identity over the domain

$$|\xi| < R, \quad |\eta| < R, \quad 0 < \tau < t - \epsilon. \quad \text{for } 0 < \epsilon < t, \tag{7.3}$$

and apply Stokes' theorem. Then, letting $R \to \infty$, $\epsilon \downarrow 0$, we have for $v(x, y)$ tending to zero (as $(x, y) \to \infty$) in an appropriate sense

$$u(x, y, t) = \int_{-\infty}^{\infty} \int_{-\infty}^{\infty} G(x, y, t; \xi, \eta, 0)v(\xi, \eta)d\xi d\eta$$

where $v(x, y)$ is the temperature at $t = 0$.

Letting $t = 1$ in the latter identity, we have

$$\int_{-\infty}^{\infty} \int_{-\infty}^{\infty} v(\xi, \eta)e^{-\frac{(x-\xi)^2+(y-\eta)^2}{4}} d\xi d\eta = 4\pi u(x, y, 1). \tag{7.4}$$

We first consider a case where the moment method can be applied. Assume that

$$\operatorname{supp} v(x, y) \subset \overline{\mathbf{R}}_+ \times \overline{\mathbf{R}}_+.$$

Then, Equation (7.4) can be written

$$\int_0^{\infty} \int_0^{\infty} v(\xi, \eta)e^{-\frac{(x-\xi)^2+(y-\eta)^2}{4}} d\xi d\eta = 4\pi u(x, y, 1), \quad (x, y) \in \mathbf{R}^2,$$

where $u(x, y, 1)$, which is analytic in x and y, is assumed to be in $L^2(\mathbf{R}^2)$. For $x = -2, -4, ..., y = -2, -4, ...$ the latter equation becomes after some rearrangements

$$e^{-(m^2+n^2)} \int_0^{\infty} \int_0^{\infty} v(\xi, \eta)e^{-\frac{\xi^2+\eta^2}{4}} e^{-(m\xi+n\eta)} d\xi d\eta = f_{mn} \tag{7.5}$$

for $m, n = 1, 2, ..., f_{mn} = 4\pi u(-2m, -2n, 1)$.

Using the new variables $(s,t) = (e^{-\xi}, e^{-\eta})$, we get after some computations

$$\int_0^1 \int_0^1 w(s,t) s^i t^j \, ds \, dt = \mu_{ij}, \qquad i,j = 0,1,2,\dots \tag{7.6}$$

where

$$w(s,t) = v(-\ln s, -\ln t) e^{-\frac{\ln^2 s + \ln^2 t}{4}},$$
$$\mu_{ij} = 4\pi e^{(i+1)^2 + (j+1)^2} u(-2i-2, -2j-2, 1).$$

The problem (7.6) is a two-dimensional Hausdorff moment problem and we can use the method in Chapter 4 or [AGT] to regularize it.

For the reader's convenience, we recall some notations. For $m, n = 0, 1, 2, \dots$ we put

$$L_m(s) = (2m+1)^{1/2} \frac{1}{m!} \frac{d^m}{ds^m}(s^m(1-s)^m),$$
$$L_{mn}(s,t) = L_m(s) L_n(t)$$

and note that (L_{mn}) is an orthonormal basis for $L^2(I)$ (where $I = (0,1) \times (0,1)$). For each real sequence $\mu = (\mu_{ij})$, $i,j = 0,1,2,\dots$, we define the sequence $\lambda = \lambda(\mu) = (\lambda_{ij})$ by

$$\lambda_{ij} = \lambda_{ij}(\mu) = \sum_{p=0}^i \sum_{q=0}^j C_{ip} C_{jq} \mu_{pq},$$

where

$$C_{mj} = (2m+1)^{1/2}(-1)^j \frac{(m+j)!}{(j!)^2(m-j)!}, \qquad 0 \le j \le m,$$

and we put

$$p^n = p^n(\mu) = \sum_{i,j=0}^n \lambda_{ij}(\mu) L_{ij},$$
$$q^n(\xi,\eta) = e^{\frac{\xi^2 + \eta^2}{4}} p^n(e^{-\xi}, e^{-\eta}).$$

Let L_ρ^2 be the space of all functions f such that

$$\sqrt{\rho}\, f \in L^2(\mathbf{R}_+ \times \mathbf{R}_+)$$

where

$$\rho(\xi,\eta) = e^{-(\xi+\eta) - \frac{\xi^2 + \eta^2}{2}}.$$

Then, we have

Theorem 7.1. *Let (f_{mn}) be a sequence of real numbers. If (7.5) has a solution v in $L^2(\mathbf{R}_+ \times \mathbf{R}_+)$ then*

$$q^n \longrightarrow v \quad \text{in } L_\rho^2.$$

Moreover, if the solution v is in $W^{1,\infty}(\mathbf{R}_+ \times \mathbf{R}_+)$ then there is a constant C independent of v and n such that

$$\|q^n - v\|_{L^2_\rho} \leq \frac{C}{2(n+1)}\|v\|_{W^{1,\infty}}$$

where (according to the definition of the space L^2_ρ)

$$\|\phi\|_{L^2_\rho} = \|\sqrt{\rho}\ \phi\|_{L^2} \qquad \forall \phi \in L^2_\rho.$$

Proof. Since $v \in L^\infty(\mathbf{R}_+ \times \mathbf{R}_+)$, the function

$$w(s,t) = v(-\ln s, -\ln t)e^{-\frac{\ln^2 s + \ln^2 t}{4}}$$

is in $L^2(I)$ and by (7.6) w satisfies a two-dimensional Hausdorff moment problem. Hence, applying Theorem 4.1, we get

$$\|p^n - w\|_{L^2(I)} \longrightarrow 0 \qquad \text{as } n \to \infty. \tag{7.7}$$

But

$$
\begin{aligned}
\|p^n - w\|^2_{L^2(I)} &= \int_0^1 \int_0^1 |p^n(s,t) - w(s,t)|^2 ds dt \\
&= \int_0^\infty \int_0^\infty \left(q^n(e^{-\xi}, e^{-\eta}) - v(\xi,\eta)e^{-\frac{\xi^2+\eta^2}{4}}\right)^2 e^{-(\xi+\eta)} d\xi d\eta \\
&= \int_0^\infty \int_0^\infty \left(q^n(e^{-\xi}, e^{-\eta}) - v(\xi,\eta)\right)^2 \rho(\xi,\eta) d\xi d\eta \\
&= \|q^n - v\|^2_{L^2_\rho}. \tag{7.8}
\end{aligned}
$$

Combining (7.7), (7.8) gives $q^n \to v$ in L^2_ρ. Now, if $v \in W^{1,\infty}(\mathbf{R}_+ \times \mathbf{R}_+)$ then one has $w \in H^1(I)$. Moreover,

$$\frac{\partial w}{\partial s} = -\left(\frac{1}{s}\frac{\partial v}{\partial \xi} + \frac{v \ln s}{2s}\right)e^{-\frac{\ln^2 s + \ln^2 t}{4}}.$$

It follows that

$$\left\|\frac{\partial w}{\partial s}\right\|_{L^2(I)} \leq C_0\|v\|_{W^{1,\infty}}.$$

Analogously,

$$\left\|\frac{\partial w}{\partial t}\right\|_{L^2(I)} \leq C_0\|v\|_{W^{1,\infty}}.$$

Hence,

$$\|w\|_{H^1(I)} \leq C\|v\|_{W^{1,\infty}}. \tag{7.9}$$

Using the result of Theorem 4.1, we get

$$\|p^n(\mu) - w\|_{L^2(I)} \leq \frac{1}{2(n+1)}\|w\|_{H^1(I)}.$$

In view of (7.8), (7.9), this inequality gives

$$\|q^n - v\|_{L^2_\rho} \leq \frac{C}{2(n+1)}\|v\|_{W^{1,\infty}}.$$

This completes the proof of Theorem 7.1.

Because for a given sequence (f_{mn}) a solution of (7.5) may not exist, the following theorem will be useful.

Theorem 7.2. Let $v_0 \in L^\infty(\mathbf{R}_+ \times \mathbf{R}_+)$ be the solution of (7.5) corresponding to $f^0 = (f^0_{mn})$ in the right hand side of (7.5). Let

$$F(\theta) = \frac{729}{320}(2\theta + 1)3^{4\theta}, \qquad \theta \geq 1$$

and for $0 < \epsilon < 1$, put

$$n(\epsilon) = [F^{-1}(\epsilon^{-1/2})]$$

where $[x]$ is the largest integer $\leq x$.
Then there exists a function $\eta(\epsilon)$, $0 < \epsilon < 1$, such that

$$\eta(\epsilon) \longrightarrow 0 \qquad as \ \epsilon \to 0$$

and that for all sequences $f = (f_{mn})$ satisfying

$$\sup_{m,n} \left| e^{m^2+n^2}(f_{mn} - f^0_{mn}) \right| < \epsilon$$

we have

$$\|q^{n(\epsilon)} - v_0\|_{L^2_\rho} \leq \eta(\epsilon).$$

Moreover, if $v_0 \in W^{1,\infty}(\mathbf{R}_+ \times \mathbf{R}_+)$ then

$$\|q^{n(\epsilon)} - v_0\|_{L^2_\rho} \leq \epsilon^{1/2} + \frac{C\|v_0\|_{W^{1,\infty}}}{C(\epsilon)}$$

with

$$C(\epsilon) = 2(2\ln 3 + \frac{1}{2}\ln 2)^{-1}\ln\left(\frac{8\sqrt{5}}{27\sqrt{2}\sqrt[4]{\epsilon}}\right).$$

Proof. By Theorem 4.2, there exists a function $\eta(\epsilon)$, $\lim_{\epsilon\downarrow 0}\eta(\epsilon) = 0$ such that

$$\|p^{n(\epsilon)}(\mu) - w_0\|_{L^2(I)} \leq \eta(\epsilon) \tag{7.10}$$

where

$$w_0(s,t) = v_0(-\ln s, -\ln t)e^{-\frac{\ln^2 s + \ln^2 t}{4}}.$$

From (7.8), (7.10) one has

$$\|q^{n(\epsilon)}(\mu) - v_0\|_{L_\rho^2} \leq \eta(\epsilon).$$

Now if $v_0 \in W^{1,\infty}(\mathbf{R}_+^2)$, then, as shown in Theorem 7.1, we have $w_0 \in H^1(0,1)$ and

$$\|w_0\|_{H^1(I)} \leq C\|v_0\|_{W^{1,\infty}}. \tag{7.11}$$

From the error estimates in Theorem 4.2, one has

$$\|p^{n(\epsilon)}(\mu) - w_0\|_{L^2(I)} \leq \epsilon^{1/2} + \frac{C\|w_0\|_{H^1(I)}}{C(\epsilon)} \tag{7.12}$$

with

$$C(\epsilon) = 2(2\ln 3 + \frac{1}{2}\ln 2)^{-1} \ln \left(\frac{8\sqrt{5}}{27\sqrt{2}\sqrt[4]{\epsilon}} \right).$$

Combining (7.11), (7.12) completes the proof of Theorem 7.2.

Now we turn to a case where the method of truncated integration can be applied. We rewrite (7.4) as

$$\int_{-\infty}^{\infty} \int_{-\infty}^{\infty} v(\xi,\eta) e^{-\frac{(x-\xi)^2+(y-\eta)^2}{4}} d\xi d\eta = g(x,y), \tag{7.13}$$

where $g(x,y) = 4\pi u(x,y,1)$. This is an integral equation (in fact, a convolution equation) in the unknown function $v(\xi,\eta)$. It will be regularized by truncated integration as follows.

Taking the Fourier transform of both sides of (7.13) gives formally

$$\hat{v}(\omega,\zeta) \, e^{-(\omega^2+\zeta^2)} = \hat{g}(\omega,\zeta), \tag{7.14}$$

where

$$\hat{h}(\omega,\zeta) = \int_{\mathbf{R}^2} h(x,y) e^{i(x\omega+y\zeta)} dx dy.$$

Note that the inverse of the Fourier transform is given by the formula

$$h(x,y) = \frac{1}{(2\pi)^2} \int_{\mathbf{R}^2} \hat{h}(\omega,\zeta) e^{-i(x\omega+y\zeta)} d\omega d\zeta.$$

From (7.14), we see that if a solution v exists, then $\hat{v}$ is given by

$$\hat{v}(\omega,\zeta) \, = e^{\omega^2+\zeta^2} \hat{g}(\omega,\zeta). \tag{7.15}$$

Thus, if a solution v of (7.13) exists (in $L^2 \equiv L^2(\mathbf{R}^2)$), then $e^{\omega^2+\zeta^2}\hat{g}$ has to be in L^2. We see that if g is an arbitrary function in L^2, then solutions may not exist. Furthermore, in the case of existence, solutions do not depend continuously on g. From (7.15), it is seen that if a solution exists, i.e., if $e^{\omega^2+\zeta^2}\hat{g}$ is in L^2, then

$$v(x,y) \, = \frac{1}{4\pi^2} \int_{\mathbf{R}^2} e^{\omega^2+\zeta^2} \hat{g}(\omega,\zeta) \, e^{-i(x\omega+y\zeta)} d\omega d\zeta. \tag{7.16}$$

We shall construct a regularized solution stable with respect to variations in g, even if a solution corresponding to the measure data g does not exist. We shall apply the method of truncated integration after applying a Fourier transform.

Now, let g_0 be such that (7.13) has a solution $v_0 \in L^2 \equiv L^2(\mathbf{R}^2)$ corresponding to g_0 in the right hand side. Let g be such that

$$\|g - g_0\|_{L^2} \le \epsilon. \tag{7.17}$$

We construct a function v stable with respect to variations in g. If v_0 is sufficiently smooth, then we can estimate the L^2-error between v_0 and v.

Our main results in the method of truncated integration are the following two theorems.

Theorem 7.3. *Let inequality (7.17) hold. Assume that*

$$\int_{\mathbf{R}^2} e^{2(\omega^2 + \zeta^2)} |\hat{g}_0(\omega, \zeta)|^2 \, (\omega^2 + \zeta^2)^2 d\omega d\zeta \le E^2. \tag{7.18}$$

Put

$$v(x, y) = \frac{1}{4\pi^2} \int_{-\pi/h}^{\pi/h} \int_{-\pi/h}^{\pi/h} e^{\omega^2 + \zeta^2} \hat{g}(\omega, \zeta) \, e^{-i(x\omega + y\zeta)} d\omega d\zeta, \tag{7.19}$$

where

$$h^2 = \frac{2\pi^2}{\ln\left\{ \dfrac{E}{\epsilon} \left(\dfrac{E}{\epsilon} \right)^{-1} \right\}}. \tag{7.20}$$

Then, for $\epsilon < E/e$, we have the error estimate

$$\|v_0 - v\|_{L^2} \le \frac{E\sqrt{2}}{\ln\left\{ \dfrac{E}{\epsilon} \left(\ln \dfrac{E}{\epsilon} \right)^{-1} \right\}}. \tag{7.21}$$

Proof. In view of Plancherel's theorem, we have

$$\|v_0 - v\|_{L^2}^2 = \frac{1}{4\pi^2} \|\hat{v}_0 - \hat{v}\|_{L^2}^2$$

$$\le \frac{1}{4\pi^2} \int_{\omega^2 + \zeta^2 \le R_\epsilon^2} e^{\omega^2 + \zeta^2} |\hat{g} - \hat{g}_0|^2 d\omega d\zeta$$

$$+ \frac{1}{4\pi^2} \int_{\omega^2 + \zeta^2 \ge R_\epsilon^2} e^{\omega^2 + \zeta^2} |\hat{g}_0|^2 d\omega d\zeta,$$

where

$$R_\epsilon^2 = \ln\left\{ \frac{E}{\epsilon} \left(\ln \frac{E}{\epsilon} \right)^{-1} \right\}.$$

Hence

$$\|v_0 - v\|_{L^2}^2 \leq \frac{1}{4\pi^2} e^{2R_\epsilon^2} \int_{\mathbf{R}^2} |\hat{g} - \hat{g}_0|^2 \, d\omega d\zeta$$

$$+ \frac{1}{4\pi^2 R_\epsilon^4} \int_{\mathbf{R}^2} e^{\omega^2 + \zeta^2} |\hat{g}_0|^2 (\omega^2 + \zeta^2)^2 \, d\omega d\zeta$$

$$\leq E^2 \left(\frac{1}{\ln^2(E/\epsilon)} + \frac{1}{R_\epsilon^4} \right).$$

For $\epsilon < E/e$ we have

$$R_\epsilon^2 = \ln \left\{ \frac{E}{\epsilon} \left(\ln \frac{E}{\epsilon} \right)^{-1} \right\} \leq \ln \frac{E}{\epsilon}.$$

Therefore

$$\|v_0 - v\|_{L^2}^2 \leq \frac{2E^2}{R_\epsilon^4} = \frac{2E^2}{\ln^2 \left\{ \frac{E}{\epsilon} \left(\ln \frac{E}{\epsilon} \right)^{-1} \right\}}.$$

This completes the proof of Theorem 7.3.

Theorem 7.4. *The function v defined as in (7.19) can be represented by the double cardinal series*

$$v(x, y) = \sum_{m,n=-\infty}^{\infty} v(mh, nh) S(m, h)(x) \ S(n, h)(y),$$

where $S(p, d)(z)$ is the Sinc function defined by

$$S(p, d)(z) = \frac{sin(\pi(z - pd)/d)}{\pi(z - pd)/d},$$

$p = 0, \pm 1, \pm 2, ...,$ $d > 0$, *the series converging in $L^2(\mathbf{R}^2)$.*

Proof. We have from (7.19)

$$\text{supp } \hat{v} \subset [-\pi/h, \pi/h].$$

Hence, using the results of Sect. 5.3, we get the desired expansion. This completes the proof of Theorem 7.4.

Remark 7.1. By Theorem 7.4, $v(x, y)$ is completely determined by its values at the lattice points (mh, nh), $m, n = 0, \pm 1, \pm 2, \ldots$.

Remark 7.2. For notational convenience, we considered only the two-dimensional case. The three-dimensional case can be treated quite analogously.

7.2 Surface temperature determination from borehole measurements: a two-dimensional problem

The problems of surface temperature determination from borehole measurements has been treated widely in the case of one space dimension (see [ASa] for the early literature, also [LN1], [LN2]). The literature in the case of two space dimensions is rather scarce. We mention the work of Kaminski and Grysa [GK] where the problem is treated numerically and the more recent works (see [Le1], [Le2] and [LNTT]). Consider the heat equation

$$\Delta u - u_t = 0, \qquad (x, y) \in \mathbf{R} \times \mathbf{R}_+, \ t > 0, \tag{7.22}$$

where the half space $\mathbf{R} \times \mathbf{R}_+$ represents the Earth, the interior of which consists of the points (x, y), $-\infty < x < \infty$, $y > 0$. In (7.22) u denotes the temperature function and the heat conductivity is, for notational convenience, taken to be 1. The problem is to determine the temperature in the half plane $-\infty < x < \infty$, $y > 0$. A natural way would be to measure the temperature on the surface $y = 0$ and compute from it the temperature at inner points. While the determination of the temperature is a stable process, measurements of surface temperature are usually contaminated with noise. Therefore, one is led to measure temperature at boreholes in the interior of the Earth, e.g., at points on the line $y = 1$ (the positive direction of y-axis is into the Earth), although the computation of the surface temperature from the temperature measured at $y = 1$ is an ill-posed problem.

We shall formulate the problem as a moment problem and then regularize it using the results of Chapter 5. Specifically, we shall determine a function $v(x, t) = u(x, 0, t)$ from a certain sequence of functions $(\mu_n(t))$, $\mu_n(t) = u(x_n, 1, t)$, (x_n), $n = 1, 2, ...$, being a given bounded sequence in $\mathbf{R}$ with $x_i \neq x_j$ for $i \neq j$.

We first formulate the problem as an integral equation in $v(\xi, \tau)$. Let

$$G(x, y, t; \xi, \eta, \tau) = \frac{1}{4\pi(t - \tau)} \left\{ \exp\left(-\frac{(x - \xi)^2 + (y - \eta)^2}{4(t - \tau)}\right) - \exp\left(-\frac{(x - \xi)^2 + (y + \eta)^2}{4(t - \tau)}\right) \right\}$$

be Green's function corresponding to Dirichlet's boundary condition from $y = 0$. Note that

$$G_{\xi\xi} + G_{\eta\eta} + G_\tau = 0.$$

Hence one has for $u = u(\xi, \eta, \tau)$

$$\text{div}(u\nabla G - G\nabla u) + (Gu)_\tau = 0.$$

Integrate this identity over the domain

$$|\xi| \le R, \ 0 < \eta < R, \ 0 < \tau < t - \epsilon, \qquad 0 < \epsilon < t,$$

and apply Stokes' theorem. Then, letting $R \to \infty$, $\epsilon \to 0$, we get, after some computations,

$$u(x, y, t) = \int_{-\infty}^{\infty} \int_0^{\infty} G(x, y, t; \xi, \eta, 0) u(\xi, \eta, 0) d\xi d\eta +$$

$$+ \frac{y}{4\pi} \int_0^t \int_{-\infty}^{\infty} \frac{v(\xi, \tau)}{(t - \tau)^2} \exp\left(-\frac{(x - \xi)^2 + y^2}{4(t - \tau)}\right) d\xi d\tau.$$

Letting $y = 1$ in the preceding relation, we get

$$\frac{1}{4\pi} \int_0^t \int_{-\infty}^{\infty} \frac{v(\xi, \tau)}{(t - \tau)^2} \exp\left(-\frac{(x - \xi)^2 + 1}{4(t - \tau)}\right) d\xi d\tau = h(x, t) \qquad (7.23)$$

where

$$h(x, t) = u(x, 1, t) - \int_{-\infty}^{\infty} \int_0^{\infty} G(x, 1, t; \xi, \eta, 0) u(\xi, \eta, 0) d\xi d\eta.$$

This is an integral equation in $v(\xi, \tau)$.

Let (x_n) be a real bounded sequence with $x_i \neq x_j$ for $i \neq j$. From (7.23) we have

$$\frac{1}{4\pi} \int_0^t \int_{-\infty}^{\infty} \frac{v(\xi, \tau)}{(t - \tau)^2} \exp\left(-\frac{(x_n - \xi)^2 + 1}{4(t - \tau)}\right) d\xi d\tau = \nu_n(t), \qquad (7.24)$$

where $t > 0$ and

$$\nu_n(t) = u(x_n, 1, t) - \int_{-\infty}^{\infty} \int_0^{\infty} G(x_n, 1, t; \xi, \eta, 0) u(\xi, \eta, 0) d\xi d\eta, \quad n = 1, 2, \dots .$$

This is a moment problem in $v(\xi, \tau)$. We shall consider the sequence of numbers

$$x_n = \frac{1}{\pi} \ln\left(1 + \frac{2}{n}\right).$$

Let $v_0 \in L^2(\mathbf{R} \times \mathbf{R}_+)$ be a solution of (7.24) corresponding to $\nu^0(t) = (\nu_n^0(t))$, $\nu_n^0 \in L^{\infty}(\mathbf{R}_+)$, in the right hand side. Let $\nu(t) = (\nu_n(t))$, $\nu_n \in L^{\infty}(\mathbf{R}_+)$, satisfy

$$\|\nu_n^0 - \nu_n\|_{L^{\infty}} < \epsilon.$$

From the sequence of functions (ν_n), we shall construct a regularized solution for (7.24). For $m \in \mathbf{N}$, we denote by $a_{mk}(t)$ $(t > 0,\ k = 0, 1, .., m - 1)$ the solution of the linear system

$$\sum_{k=0}^{m-1} a_{mk}(t) \omega_n^k = \nu_n(t), \qquad n = 1, 2, .., m,$$

where $\omega_n = 1/(n + 1)$, $n = 1, 2, \dots$. By the classical theory of polynomial interpolation, the functions $a_{mk}(t)$ exist uniquely. Put

$$P_m(\omega, t) = \sum_{0 \leq k \leq m/2} a_{mk}(t) \omega^k, \qquad t > 0, \qquad (7.25)$$

and

$$h^*(x,t) = P_m\left(\frac{e^{\pi x} - 1}{e^{\pi x} + 1}, t\right), \qquad \text{for } |x|, |t| \le \frac{1}{2\pi}\ln(m),$$
$$= 0 \quad otherwise. \tag{7.26}$$

For every $\beta > 0$, we let

$$v_{\beta m}(x,t) = \frac{1}{4\pi^2} \int_{\mathbf{R}^2} \Psi(\xi,\eta) e^{-i(x\xi + y\eta)} \, d\xi \, d\eta, \tag{7.27}$$

where

$$\Psi = \bar{\hat{a}}\hat{h}^*(\beta + |\hat{a}|^2)^{-1},$$
$$\alpha(x,t) = \frac{1}{t^2}\exp\left(-\frac{x^2+1}{4t}\right) \quad \text{for } t > 0,$$
$$= 0 \qquad \text{for } t < 0.$$

We shall prove that there are functions $\beta(\epsilon)$, $m(\epsilon)$ such that $(v_{\beta(\epsilon)m(\epsilon)})$ is a family of regularized solutions. In fact, we have

Theorem 7.5. *Suppose the exact solution v_0 of (7.24), corresponding to $\nu^0(t) = (\nu_n^0(t))$ $(\nu_n^0 \in L^\infty(\mathbf{R}_+))$ in the right hand side, satisfies*

$$v_0 \in L^\infty(\mathbf{R} \times \mathbf{R}_+), \quad (1 + x^2 + t^2)^2 v_0 \in L^2(\mathbf{R} \times \mathbf{R}_+).$$

Let $\nu(t) = (\nu_n(t))$, $\nu_n \in L^\infty(\mathbf{R}_+)$, satisfy

$$\sup_n \|\nu_n - \nu_n^0\|_{L^\infty(\mathbf{R}_+)} < \epsilon.$$

Put

$$g(\theta) = \left(\frac{\theta}{2} + 1\right)\left(\frac{4(\theta+1)\sqrt{2e}}{\sqrt{\theta}}\right)^\theta, \qquad \theta > 0,$$
$$m(\epsilon) = [g^{-1}(\epsilon^{-1/2})], \qquad \epsilon > 0,$$
$$\beta^2(\epsilon) = \left\{\epsilon^{1/2} + 4\pi^2\left(\frac{m(\epsilon)}{2} + 1\right)\left(\frac{4\sqrt{2e}}{m(\epsilon)}\right)^{m(\epsilon)}\right.$$
$$\left. + \left(1 - \frac{1}{\sqrt{m(\epsilon)}}\right)^{m(\epsilon)/2}\sqrt{m(\epsilon)}\right\}\ln(m(\epsilon)) + \frac{1}{\ln(m(\epsilon))},$$
$$v_\epsilon = v_{\beta(\epsilon)m(\epsilon)}, \qquad v_{\beta m} \text{ as in (7.27).}$$

Then there exists a $C > 0$ depending only on v_0 such that

$$\|v_\epsilon - v_0\|_{L^2(\mathbf{R} \times \mathbf{R}_+)} \le C\left(\ln\left(\frac{1}{\beta(\epsilon)}\right)\right)^{-1}.$$

Proof. Put

$$h_0(z,t) = \frac{1}{4\pi} \int_0^t \int_{-\infty}^{\infty} \frac{v_0(\xi,\tau)}{(t-\tau)^2} \exp\left(-\frac{(z-\xi)^2+1}{4(t-\tau)}\right) d\xi d\tau.$$

The remainder of the proof is divided into two parts. In the first parts (consisting of Step 1 and Step 2), we shall estimate $\|h_0 - h^*\|_{L^2(\mathbf{R}\times\mathbf{R}_+)}$. In the second part (Step 3), we shall derive an estimate for $\|v_\epsilon - v_0\|_{L^2(\mathbf{R}\times\mathbf{R}_+)}$.

From the definition of h_0, it follows that $h_0(.,t)$ is, for each $t > 0$, analytic in the strip $|Im\ z| < 1$ of the complex plane.

For every z satisfying $|Im\ z| < 1/2$ we put

$$\omega = T(z) = \frac{e^{\pi z} - 1}{e^{\pi z} + 1}.$$

We can verify directly that T is a homeomorphism of the strip

$$\{z \in \mathbf{C}:\ |Im\ z| < 1/2\}$$

onto the open unit disc U. Put

$$\tilde{h}(\omega,t) = h_0(T^{-1}\omega, t).$$

Following are three steps of the proof.

Step 1. *Estimate of $|\tilde{h}(\omega,t) - P_m(\omega,t)|,\ \ \omega \in U$:*

Since $v_0 \in L^\infty(\mathbf{R} \times \mathbf{R}_+)$, one has

$$
\begin{aligned}
|\tilde{h}(\omega,t)| &= |h(z,t)| \\
&\leq \frac{\|v_0\|_{L^\infty}}{4\pi} \int_0^t \int_{-\infty}^{\infty} \frac{1}{(t-\tau)^2} \exp\left(-\frac{(x-\xi)^2+1}{4(t-\tau)}\right) d\xi d\tau \\
&\leq C\|v_0\|_{L^\infty}, \qquad \forall \omega \in \bar{U}
\end{aligned}
\tag{7.28}
$$

where

$$C = \sup_{(x,t)\in\mathbf{R}\times\mathbf{R}_+} \frac{1}{4\pi} \int_0^\infty \int_{-\infty}^{\infty} \frac{1}{(t-\tau)^2} \exp\left(-\frac{(x-\xi)^2+1}{4(t-\tau)}\right) d\xi d\tau.$$

On the other hand, in view of the analyticity of $\tilde{h}$ in the variable ω one has

$$\tilde{h}(\omega,t) = \sum_{n=0}^{\infty} a_n(t)\omega^n \tag{7.29}$$

where

$$a_n(t) = \frac{1}{2\pi i} \int_{\partial U} \xi^{-n-1}\tilde{h}(\xi,t)d\xi. \tag{7.30}$$

We have

$$\tilde{h}(\omega_n, t) = h_0(x_n, t) = v_n^0(t). \tag{7.31}$$

Using the calculations in the proof of Theorem 5.1, Section 5.1, one has in view of (7.29), (7.31) the equation

$$a_{mk}(t) - a_k(t) \equiv c_{mk}(t) = \sum_{n=1}^{m} \varphi_{mn}(t) s_{mn}^{-1} \sigma_{m-k-1}(\hat{\omega}_n)(-1)^{m-k-1}.$$

Here, we recall,

$$\varphi_{mn}(t) = \nu_n(t) - \nu_n^0(t) + \sum_{k=m}^{\infty} a_k(t)\omega_n^k,$$

$$s_{mn} = (\omega_n - \omega_1)...(\omega_n - \omega_{n-1})(\omega_n - \omega_{n+1})...(\omega_n - \omega_m),$$

$$\sigma_{mn}(\tilde{t}) = \sum_{1 \leq j_1 < j_2 < ... < j_k \leq m-1} t_{j_1}...t_{j_k}$$

with $\tilde{t} = (t_1, ..., t_{m-1})$. Especially, one has

$$|\varphi_{mn}(t)| \leq \epsilon + \sup_k \|a_k\|_{L^\infty} \cdot \frac{|\omega_n|^m}{1 - |\omega_n|}. \tag{7.32}$$

From (7.28), (7.30), one gets

$$\|a_k\|_{L^\infty} \leq C\|v_0\|_{L^\infty}. \tag{7.33}$$

Hence, using inequality (5.23) in Chap. 5, one gets, in view of (7.32), (7.33), the estimate

$$|a_{mk}(t) - a_k(t)| \leq \epsilon \left(\frac{4(m+1)\sqrt{2e}}{\sqrt{m}}\right)^m + 4C\|v_0\|_{L^\infty} \left(\frac{4\sqrt{2e}}{\sqrt{m}}\right)^m,$$
$$k = 0, 1, ..., m-1. \tag{7.34}$$

Now, we have

$$\tilde{h}(\omega, t) - P_m(\omega, t) = \sum_{0 \leq k \leq m/2} c_{mk}(t)\omega^k + \sum_{k > m/2} a_k(t)\omega^k.$$

Hence, for $|\omega| \leq 1 - \delta$, $0 < \delta < 1$, one gets in view of (7.34,

$$|\tilde{h}(\omega, t) - P_m(\omega, t)|$$
$$\leq \sum_{0 \leq k \leq m/2} |c_{mk}(t)| + (1-\delta)^{m/2} \sum_{k > m/2} |a_k(t)| \cdot |\omega|^{k-m/2}$$
$$\leq (1 + m/2)\left\{\epsilon \left(\frac{4(m+1)\sqrt{2e}}{\sqrt{m}}\right)^m + 4C\|v_0\|_{L^\infty} \left(\frac{4\sqrt{2e}}{\sqrt{m}}\right)^m\right\}$$
$$+ (1-\delta)^{m/2} C\|v_0\|_{L^\infty} \cdot \frac{1}{\delta}. \tag{7.35}$$

Put

$$g(t) = \left(\frac{t}{2} + 1\right)\left(\frac{4(t+1)\sqrt{2e}}{\sqrt{t}}\right)^t,$$

$$m(\epsilon) = [g^{-1}(\epsilon^{-1/2})], \ \delta = \frac{1}{\sqrt{m(\epsilon)}}.$$

Using (7.35) we obtain

$$|\tilde{h} - h^*_{m(\epsilon)}| \le \gamma(\epsilon), \qquad \forall |\omega| < 1 - \frac{1}{\sqrt{m(\epsilon)}}, \tag{7.36}$$

where

$$\gamma(\epsilon) = \epsilon^{1/2} + 8\pi C \|v_0\|_{L^\infty} \left(\frac{m(\epsilon)}{2} + 1\right)\left(\frac{4\sqrt{2e}}{\sqrt{m(\epsilon)}}\right)^{m(\epsilon)}$$

$$+ C\|v_0\|_{L^\infty}\sqrt{m(\epsilon)}(1 - \frac{1}{\sqrt{m(\epsilon)}})^{m(\epsilon)/2}.$$

Step 2. *Estimate of* $\|h_0 - h^*_{m(\epsilon)}\|_{L^2(\mathbf{R}\times\mathbf{R}_+)}$:

For convenience, we extend h_0 on $\mathbf{R}^2$ by putting $h_0(x,t) = 0$ for $t < 0$. One has

$$\int_{\mathbf{R}^2} |h_0(x,t) - h^*_{m(\epsilon)}(x,t)|^2 dx dt = I_1 + I_2, \tag{7.37}$$

where

$$I_1 = \int_{Q_\epsilon} |h_0(x,t) - h^*_{m(\epsilon)}(x,t)|^2 dx dt,$$

$$I_2 = \int_{\mathbf{R}^2 \setminus Q_\epsilon} |h_0(x,t)|^2 dx dt,$$

and Q_ϵ is the rectangle

$$Q_\epsilon = [-\frac{1}{2\pi}\ln(m(\epsilon)), \frac{1}{2\pi}\ln(m(\epsilon))] \times [-\ln(m(\epsilon)), \ln(m(\epsilon))].$$

From (7.36) one has

$$|I_1| \le \frac{1}{\pi}\gamma^2(\epsilon)\ln(m(\epsilon)). \tag{7.38}$$

On the other hand, using Schwarz's inequality one has

$$|h_0(x,t)|^2 \le \|(1 + \xi^2 + \tau^2)^2 v_0\|_{L^2}^2 \times J, \tag{7.39}$$

where

$$J = \int_0^t \int_{-\infty}^\infty \frac{1}{(1 + \xi^2 + \tau^2)^4 (t - \tau)^4} \exp\left(-\frac{(x - \xi)^2 + 1}{2(t - \tau)}\right) d\xi d\tau$$

Using the inequality $e^y \ge y^2/2$ for $y \ge 0$, one gets

$$J = \int_0^t \int_{-\infty}^{\infty} \frac{4}{(1+\xi^2+\tau^2)^4(t-\tau)^2} \times$$

$$\frac{1}{((x-\xi)^2+1/2)^2} \exp\left(-\frac{1}{4(t-\tau)}\right) d\xi d\tau$$

$$\leq 4 \int_0^t \frac{1}{(1+\tau^2)^4(t-\tau)^2} \exp\left(-\frac{1}{4(t-\tau)}\right) d\tau \times$$

$$\times \int_{-\infty}^{\infty} \frac{d\xi}{(1+\xi^2)^4((x-\xi)^2+1/2)^2}. \tag{7.40}$$

From (7.39), (7.40) one has, after some computations

$$|h_0(x,t)|^2 \leq \frac{C}{(1+t^2)(1+x^2)}$$

for some constant $C > 0$.

Hence,

$$I_2 \equiv \int_{\mathbf{R}^2 \setminus Q_\epsilon} |h_0(x,t)|^2 \, dx dt \leq \frac{C'}{\ln(m(\epsilon))}. \tag{7.41}$$

Substituting (7.38), (7.41) into (7.37), we get

$$\|h_0 - h^*\|_{L^2}^2 \leq \frac{1}{\pi}\gamma^2(\epsilon)\ln(m(\epsilon)) + \frac{C'}{\ln(m(\epsilon))}.$$

Step 3. *Estimate of* $\|v_{\beta(\epsilon)m(\epsilon)} - v_0\|_{L^2(\mathbf{R}\times\mathbf{R}_+)}$:

We shall make using of the following lemma (cf [APT3]).

Lemma 7.1. *Suppose that* $w_0 \in H^1(\mathbf{R}^2) \cap L^1(\mathbf{R}^2)$ *and* $F_0 \in L^2(\mathbf{R}^2)$ *satisfy the convolution equation*

$$\alpha_k * w_0 = F_0,$$

where

$$\alpha_k(x,t) = \frac{1}{t^2} \exp\left(-\frac{x^2+k^2}{4t}\right), \qquad for\ t > 0,$$

$$= 0 \qquad for\ t < 0,$$

$$\alpha_k * w_0(x,t) = \int_{\mathbf{R}^2} \alpha_k(x-\xi, t-\tau)w_0(\xi,\tau)d\xi d\tau.$$

Let $\beta > 0$ *and let* $F \in L^2(\mathbf{R}^2)$ *satisfy*

$$\|F - F_0\|_{L^2} < \beta.$$

Put

$$w_\beta(x,t) = \frac{1}{4\pi^2} \int_{\mathbf{R}^2} \frac{\overline{\hat{\alpha}_k(\omega,\zeta)}\hat{F}(\omega,\zeta)e^{-i(x\omega+t\zeta)}dwd\zeta}{\beta + |\hat{\alpha}_k(\omega,\zeta)|^2}.$$

Then there exists a constant C *depending only on* w_0 *such that*

$$\|w_\beta - w_0\|_{L^2} \leq \frac{C}{\ln\left(\frac{1}{\beta}\right)}.$$

Moreover, if $\hat{w}_0/\hat{\alpha}_k \in L^2(\mathbf{R}^2)$ then there exists a C depending only on w_0 such that

$$\|w_\beta - w_0\|_{L^2} \leq C\sqrt{\beta}.$$

Proof. We first calculate the Fourier transform of α_k. One has

$$\hat{\alpha}_k(\omega, \zeta) = \int_{\mathbf{R}^2} \alpha_k(x, t)e^{i(x\omega + t\zeta)}\,dx dt$$

$$= \int_0^t \frac{1}{t^2}e^{it\zeta}e^{-\frac{k^2}{4t}}\,dt \int_{-\infty}^{\infty} e^{ix\omega}e^{-\frac{x^2}{4t}}\,dx.$$

Letting $\theta = x/\sqrt{t}$ we obtain

$$\int_{-\infty}^{\infty} e^{ix\omega}e^{-\frac{x^2}{4t}}\,dx = 2\sqrt{\pi t}e^{-\omega^2 t}$$

(since the Fourier transform of $e^{-u^2/2}$ is $\sqrt{2\pi}e^{-x^2/2}$). Then from the formula for the Fourier transform of $e^{-k^2/4t}t^{-3/2}$ (cf [Ed]) we get

$$\hat{\alpha}_k(\omega, \zeta) = \frac{4\pi}{k}\exp\left(-k(\omega^2 + i\zeta)^{1/2}\right).$$

By direct computation one has

$$|\hat{\alpha}_k(\omega, \zeta)| = \frac{4\pi}{k}\exp\left(-k\gamma(\omega, \zeta)\right),$$

$$\gamma(\omega, \zeta) = \left(\frac{\omega^2 + (\omega^4 + \zeta^2)^{1/2}}{2}\right)^{1/2}.$$

Since w_0 satisfies $\alpha_k * w_0 = F_0$, we have

$$\hat{\alpha}_k\hat{w}_0 = \hat{F}_0. \tag{7.42}$$

By the definition of w_β, one has

$$\beta\hat{w}_\beta + |\hat{\alpha}_k|^2\hat{w}_\beta = \overline{\hat{\alpha}_k}\hat{F}. \tag{7.43}$$

The relations (7.42), (7.43) imply that

$$\beta(\hat{w}_\beta - \hat{w}_0) + |\hat{\alpha}_k|^2(\hat{w}_\beta - \hat{w}_0) = -\beta\hat{w}_0 + \overline{\hat{\alpha}_k}(\hat{F} - \hat{F}_0). \tag{7.44}$$

Multiplying both sides of (7.44) by the conjugate of $\hat{w}_\beta - \hat{w}_0$ and integrating on $\mathbf{R}^2$, we get after some rearrangements

$$\beta\|\hat{w}_\beta - \hat{w}_0\|_{L^2}^2 + \|\hat{\alpha}_k(\hat{w}_\beta - \hat{w}_0)\|_{L^2}^2 \leq \beta(\|\hat{w}_0\|_{L^2}^2 + 1). \tag{7.45}$$

Similarly, multiplying both sides of (7.44) by the conjugate of $(\omega^2 + \zeta^2)(\hat{w}_\beta - \hat{w}_0)$ and then integrating on $\mathbf{R}^2$, we find after some computations the estimate

$$\|\sqrt{\omega^2 + \zeta^2}(\hat{w}_\beta - \hat{w}_0)\|_{L^2} \leq \|\sqrt{\omega^2 + \zeta^2}\hat{\alpha}_k\|_{L^\infty} + \|\sqrt{\omega^2 + \zeta^2}\hat{w}_0\|_{L^2}. \qquad (7.46)$$

Put

$$A = \max\{2\pi\|\sqrt{\omega^2 + \zeta^2}\hat{\alpha}_k\|_{L^\infty} + \|w_0\|_{H^1}, (\|\hat{w}_0\|_{L^2}^2 + 1)^{1/2}\}.$$

Then (7.45), (7.46) imply

$$\|\hat{\alpha}_k(\hat{w}_\beta - \hat{w}_0)\|_{L^2}^2 \leq \beta A^2, \quad \|\sqrt{\omega^2 + \zeta^2}(\hat{w}_\beta - \hat{w}_0)\|_{L^2}^2 \leq A^2. \qquad (7.47)$$

For any $\beta > 1$, let

$$D_\beta = \{(\omega, \zeta) : |\omega| \leq a_\beta, |\zeta| \leq a_\beta^2\}.$$

For $(\omega, \zeta) \in D_\beta$, one has

$$|\hat{\alpha}_k(\omega, \zeta)| \geq |\hat{\alpha}_k(a_\beta, a_\beta^2)| = \frac{4\pi}{k} \exp(-k_1 a_\beta)$$

with

$$k_1 = k\left(\frac{1}{2} + 2^{-1/2}\right)^{1/2}.$$

Hence, we get in view of (7.47)

$$\int_{D_\beta} |\hat{w}_\beta - \hat{w}_0|^2 d\omega d\zeta \leq \frac{ke^{k_1 a_\beta}}{4\pi} \int_{\mathbf{R}^2} |\hat{\alpha}(\hat{w}_\beta - \hat{w}_0)|^2 d\omega d\zeta$$

$$\leq \frac{ke^{k_1 a_\beta} A^2 \beta}{4\pi}. \qquad (7.48)$$

For $(\omega, \zeta) \notin D_\beta$ one has $\omega^2 + \zeta^2 \geq a_\beta^2$. Hence, we get in view of (7.47)

$$\int_{\mathbf{R}^2 \backslash D_\beta} |\hat{w}_\beta - \hat{w}_0|^2 d\omega d\zeta \leq \frac{1}{a_\beta^2} \int_{\mathbf{R}^2 \backslash D_\beta} (\omega^2 + \zeta^2)|\hat{w}_\beta - \hat{w}_0|^2 d\omega d\zeta$$

$$\leq \frac{A^2}{a_\beta^2}. \qquad (7.49)$$

From (7.48), (7.49) one has

$$\|w_\beta - w_0\|_{L^2}^2 = \frac{1}{4\pi^2}\|\hat{w}_\beta - \hat{w}_0\|_{L^2}^2$$

$$\leq \frac{ke^{k_1 a_\beta} A^2 \beta}{16\pi^3} + \frac{A^2}{4\pi^2 a_\beta^2}. \qquad (7.50)$$

Let a_β be a positive solution of the equation

$$\frac{ke^{k_1 a_\beta} a_\beta^2}{4\pi} = \frac{1}{\beta}.$$

One has

$$2 \ln a_\beta + k_1 a_\beta + \ln(k/4\pi) = \ln(1/\beta). \tag{7.51}$$

It follows that $a_\beta \to +\infty$ as $\beta \to 0$. Moreover, for β sufficiently small, we have

$$(2 + k_1)a_\beta \geq \ln(1/\beta), \tag{7.52}$$

hence,

$$\frac{1}{a_\beta^2} \leq \frac{(k_1 + 2)^2}{(\ln(1/\beta))^2}. \tag{7.53}$$

From (7.52), (7.53) we deduce

$$\|w_\beta - w_0\|_{L^2}^2 \leq \frac{2A^2}{4\pi^2 a_\beta^2} \leq \frac{C^2}{(\ln \frac{1}{\beta})^2},$$

where $C = \frac{(2+k_1)\sqrt{2}A}{2\pi^2}$.

Now, if we assume $\hat{w}_0/\hat{\alpha}_k \in L^2(\mathbf{R}^2)$, then by multiplying both sides of (7.44) by the conjugate of $\hat{w}_\beta - \hat{w}_0$ and then integrating over $\mathbf{R}^2$, we get after some rearrangements

$$\beta\|\hat{w}_\beta - \hat{w}_0\|_{L^2}^2 + \|\hat{\alpha}_k(\hat{w}_\beta - \hat{w}_0)\|_{L^2}^2$$
$$\leq \beta(1 + \|\hat{w}_0/\hat{\alpha}_k\|_{L^2})\|\hat{\alpha}_k(\hat{w}_\beta - \hat{w}_0)\|_{L^2}^2,$$

hence

$$\|\hat{w}_\beta - \hat{w}_0\|_{L^2}^2 \leq \beta(1 + \|\hat{w}_0/\hat{\alpha}_k\|_{L^2}).$$

This completes the proof of Lemma 7.1.

Completion of the proof of Theorem 7.5.:

Using Lemma 7.1 with $\beta = \beta(\epsilon)$, $F_0 = h_0$, $F = h_{m(\epsilon)}^*$, $\alpha_k = \alpha$ (i.e. $k = 1$), we find in view of the estimate

$$\|F_0 - F\|_{L^2} = \|h_0 - h_{m(\epsilon)}^*\|_{L^2} \leq C\beta(\epsilon),$$

(see Step 2 of the proof) that

$$\|v_\epsilon - v_0\|_{L^2} \leq \frac{C}{\ln \frac{1}{\beta(\epsilon)}}.$$

This completes the proof of Theorem 7.5.

7.3 An inverse two-dimensional Stefan problem: identification of boundary values

The Stefan problem is one of finding functions $u(x, y, t), z(x, t)$ such that

$$\Delta u - u_t = 0, \quad t > 0, \ x \in \mathbf{R}, \ 0 < y < z(x, t), \tag{7.54}$$

and u satisfies the following boundary and initial conditions

$$u(x, z(x,t), t) = 0, \qquad x \in \mathbf{R}, \ t > 0, \tag{7.55}$$

$$u(x, 0, t) = v(x, t), \qquad x \in \mathbf{R}, \ t > 0, \tag{7.56}$$

$$\frac{\partial u}{\partial n}(x, z(x,t), t) = z_t(x,t), \qquad x \in \mathbf{R}, \ t > 0, \tag{7.57}$$

$$z(x, 0) = b(x) > 0, \qquad x \in \mathbf{R}, \tag{7.58}$$

$$u(x, y, 0) = u_0(x, y), \qquad x \in \mathbf{R}, \ 0 < y < b(x). \tag{7.59}$$

The function u is usually the temperature, the zone $0 < y < z(x,t)$ is the liquid zone and the zone $y > z(x,t)$ is the ice zone. Given an initial temperature u_0, for a prescribed surface $z(x,t)$, the problem of determining the boundary values $u(x,0,t)$, i.e., the time-dependent surface temperature $v(x,t)$ is called an inverse Stefan problem. While the one-dimensional inverse Stefan problem has been treated widely (cf. [APT1], [APT2], [Bar], [Co1], [Co2], [Co3], [Jo1], [Jo2], [Wa]), the literature of the two-dimensional inverse Stefan is rather scarce. We single out the work by Colton ([Co1], [Co2]) where the given boundary $y = z(x,t)$ is assumed to be an analytic surface. In the work of Colton (loc. cit.), it is assumed that a solution exists. Now, the problem is known to be ill-posed , i.e., solutions do not always exist and whenever they do exist, there is no continuous dependence on the given data (which are here the given initial condition and free surface). We shall take the approach followed in [GAT], [APT1]-[APT4] where no solution is assumed to exist. In fact, in this section, we consider the problem of finding $v(x,t)$ from the given sequences of values $z(nh, t)$, $z_x(nh, t)$, $b(nh)$, $\tilde{u}_0(nh, mh)$ $(h > 0, m, n \in \mathbf{Z})$, where $\tilde{u}_0$ is an appropriate extension of u_0 from the set

$$\{(x, y) : \ x \in \mathbf{R}, \ 0 < y < b(x)\}$$

to $\mathbf{R}^2$.

To regularize the problem (7.54)-(7.59) we shall transform it into an integral equation. Let

$$K(x, y, t; \xi, \eta, \tau) = \frac{1}{4\pi(t - \tau)} exp\left(-\frac{(x - \xi)^2 + (y - \eta)^2}{4(t - \tau)}\right)$$

$$G(x, y, t; \xi, \eta, \tau) = K(x, y, t; \xi, \eta, \tau) + K(x, -y, t; \xi, \eta, \tau).$$

In (7.57), the normal derivative $\partial u / \partial n$ to the surface $y = z(x,t)$ can be written as

$$\frac{\partial u}{\partial n} = -\frac{\partial u}{\partial x} \frac{z_x(x,t)}{\sqrt{1 + |z_x(x,t)|^2}} + \frac{\partial u}{\partial y} \frac{1}{\sqrt{1 + |z_x(x,t)|^2}}.$$

for $y = z(x,t)$.

Let $x \in \mathbf{R}, 0 < y < z(x,t)$ and let u be sufficiently regular. The functions $u(\xi, \eta, \tau)$ and $G(.; \xi, \eta, \tau)$ satisfy the identity

$$\text{div}(u\nabla G - G\nabla u) + (Gu)_\tau = 0. \tag{7.60}$$

Integrating (7.60) over the domain $-n < \xi < n$, $0 < \eta < z(\xi, \tau)$, $1/n < \tau < t-1/n$, letting $n \to \infty$ and taking into account the initial and boundary values (7.55)-(7.59) we get an equation in $v(\xi, \tau)$, namely

$$\frac{1}{4\pi} \int_0^t \int_{-\infty}^{\infty} \frac{y}{(t-\tau)^2} \exp\left(-\frac{(x-\xi)^2 + y^2}{4(t-\tau)}\right) v(\xi, \tau) d\xi d\tau = g(x, y, t) \qquad (7.61)$$

where $t > 0, x \in \mathbf{R}, 0 < y < z(x, t)$ and $g(x, y, t)$ is defined by

$$g(x, y, t) = u(x, y, t) - \int_{-\infty}^{\infty} \int_0^{b(\xi)} u_0(\xi, \tau) G(x, y, t; \xi, \eta, 0) d\xi d\eta$$

$$\int_{-\infty}^{\infty} \int_0^t z_\tau(\xi, \tau) G(x, y, t; \xi, z(\xi, \tau), \tau) d\xi d\tau \qquad (7.62)$$

for $t > 0, x \in \mathbf{R}, 0 < y < z(x, t)$. Letting $y \uparrow z(x, t)$ in (7.61) we have

$$\frac{1}{4\pi} \int_0^t \int_{-\infty}^{\infty} \frac{z(x, t)}{(t-\tau)^2} \exp\left(-\frac{(x-\xi)^2 + z^2(x, t)}{4(t-\tau)}\right) v(\xi, \tau) d\xi d\tau$$

$$= -\int_{-\infty}^{\infty} \int_0^{b(\xi)} u_0(\xi, \tau) G(x, z(x, t), t; \xi, \eta, 0) d\xi d\eta$$

$$- \int_{-\infty}^{\infty} \int_0^t z_\tau(\xi, \tau) G(x, z(x, t), t; \xi, z(\xi, \tau), \tau) d\xi d\tau. \qquad (7.63)$$

This is an integral equation of first kind for the unknown function $v(\xi, \tau)$ and hence is ill-posed in customary function spaces. We shall transform it into an equation of convolution type for which error estimates for regularized solutions are readily derived from Lemma 7.1 of the preceding section.

Put

$$U_0(x, t) = \lim_{y \uparrow z(x, t)} g(x, y, t), \qquad (7.64)$$

$$U_1(x, t) = \frac{1}{\sqrt{1 + |z_x(x, t)|^2}} \lim_{y \uparrow z(x, t)} \left(\frac{\partial g}{\partial x} z_x(x, t) - \frac{\partial g}{\partial y}\right), \qquad (7.65)$$

and consider the function

$$U(x, y, t) = \frac{1}{4\pi} \int_0^t \int_{-\infty}^{\infty} \frac{y}{(t-\tau)^2} \exp\left(-\frac{(x-\xi)^2 + y^2}{4(t-\tau)}\right) v(\xi, \tau) d\xi d\tau.$$

Then U satisfies the equation

$$\Delta u - u_t = 0, \qquad t > 0, \ x \in \mathbf{R}, \ y > 0, \qquad (7.66)$$

and the initial and boundary conditions

$$U(x, y, 0) = 0, \qquad x \in \mathbf{R}, y > 0, t > 0,$$

$$U(x, z(x, t), t) = U_0(x, t), \qquad x \in \mathbf{R}, y > 0, t > 0, \qquad (7.67)$$

$$-\frac{\partial U}{\partial n}(x, z(x, t), t) = U_1(x, t), \qquad x \in \mathbf{R}, y > 0, t > 0. \qquad (7.68)$$

Hence $U(x, y, t)$ can be represented in terms of U_0 and U_1 on the domain $x \in \mathbf{R}, \ y > z(x, t), \ t > 0$. Let k be a number such that $k > z(x, t), \forall x \in \mathbf{R}, t > 0$. Then $U(x, k, t)$ is known, say

$$U(x, k, t) = F(x, t; z, u_0, b)$$

where $F(.; z, u_0, b)$ is calculated in terms of z, u_0, b. Thus, we arrive at the integral equation

$$\frac{1}{4\pi} \int_0^t \int_{-\infty}^\infty \frac{k}{(t - \tau)^2} \exp\left(-\frac{(x - \xi)^2 + k^2}{4(t - \tau)}\right) v(\xi, \tau) d\xi d\tau =$$

$$= F(x, t; z, u_0, b) \tag{7.69}$$

which is of the form

$$\alpha_k * v = F. \tag{7.70}$$

Here, we recall,

$$\alpha_k = \frac{1}{t^2} \exp\left(-\frac{x^2 + k^2}{4t}\right), \quad t > 0,$$

$$= 0, \quad t < 0.$$

Hence, we can use Lemma 7.1 in Section 7.2 to regularize our problem. To go into the details of the transformation from (7.67) to (7.69), we first note that (7.63), (7.64) imply

$$U_0(x, t) = -\int_{-\infty}^\infty \int_0^{b(\xi)} u_0(\xi, \tau) G(x, z(x, t), t; \xi, \eta, 0) d\xi d\eta$$

$$- \int_{-\infty}^\infty \int_0^t z_\tau(\xi, \tau) G(x, z(x, t), t; \xi, z(\xi, \tau), \tau) d\xi d\tau. \tag{7.71}$$

Furthermore, taking the normal derivative of the right hand side of (7.62) and using the jump relation (see, e.g. [Fr], Chap. 5, page 137) we have

$$U_1(x, t) = -\frac{3}{2} z_\tau(x, t)$$

$$- \int_{-\infty}^\infty \int_0^{b(\xi)} u_0(\xi, \tau) \frac{\partial G}{\partial n}(x, z(x, t), t; \xi, \eta, 0) d\xi d\eta$$

$$- \int_{-\infty}^\infty \int_0^t z_\tau(\xi, \tau) \frac{\partial G}{\partial n}(x, z(x, t), t; \xi, z(\xi, \tau), \tau) d\xi d\tau. \tag{7.72}$$

By (7.71), (7.72) the functions U_0, U_1 are calculated from z, u_0, b.
   Integrating the identity

$$\text{div}(U \nabla K - K \nabla U) + (UK)_\tau = 0$$

over the domain

$$-n < \xi < n, \quad z(\xi, \tau) < \eta < n, \quad 1/n < \tau < t - 1/n,$$

taking account of the initial and boundary values (7.67) and letting $n \to \infty$, we get

$$U(x,y,t) = \int_{-\infty}^{\infty} \int_0^t U_1(\xi,\tau) K(x,y,t;\xi,z(\xi,\tau),\tau) d\xi d\tau$$
$$- \int_{-\infty}^{\infty} \int_0^t U_0(\xi,\tau) K_1(x,y,t;\xi,z(\xi,\tau),\tau) d\xi d\tau, \qquad (7.73)$$

where

$$K_1(x,y,t;\xi,\eta,\tau) = \frac{\partial K}{\partial \xi}(x,y,t;\xi,\eta,\tau) z_\xi(\xi,\eta) -$$
$$- \frac{\partial K}{\partial \eta}(x,y,t;\xi,\eta,\tau) + K(x,y,t;\xi,\eta,\tau). \qquad (7.74)$$

Letting $y = k$ in (7.73). (7.74), we get (7.69) with

$$F(x,t;z,u_0,b) = \int_{-\infty}^{\infty} \int_0^t U_1(\xi,\tau) K(x,k,t;\xi,z(\xi,\tau),\tau) d\xi d\tau$$
$$- \int_{-\infty}^{\infty} \int_0^t U_0(\xi,\tau) K_1(x,k,t;\xi,z(\xi,\tau),\tau) d\xi d\tau. \qquad (7.75)$$

To regularize (7.69) we shall use Lemma 7.1. However, since the functions z, u_0, b are defined only on a discrete set of points, the function $F(.;z,u_0,b)$ is not known exactly. Using Sinc series, we can, under some smoothness assumptions on z, u_0, b, construct functions z_h, z_{xh}, u_{0h}, b_h approximating z, z_x, u_0, b in an appropriate sense. As in Lemma 6.1, we can construct a function $F_h(x,t;z_h,z_{xh},u_{0h},b_h)$ approximating $F(x,t;z,u_0,b)$ in the L^2-sense. In fact, for $h > 0$, we shall assume that $(\zeta_n(t))$, $(\tilde{\zeta}_n(t))$, (β_n), (ν_{mn}) $(m,n \in \mathbf{Z})$ are sequences such that $\zeta, \tilde{\zeta} \in L^2(\mathbf{R}_+)$ and that

$$\sum_{n \in \mathbf{Z}} \left(\|z(nh,.) - \zeta_n\|_{L^2(\mathbf{R}_+)}^2 + \|z_x(nh,.) - \tilde{\zeta}_n\|_{L^2(\mathbf{R}_+)}^2 \right) +$$
$$\sum_{n \in \mathbf{Z}} |b(nh) - \beta_n|^2 + \sum_{m,n \in \mathbf{Z}} |u_0(mh,nh) - \nu_{mn}|^2 < Ch.$$

From the results in Section 5.3, we get the functions approximating z, z_x, b, u_0

$$z_h(x,t) = \sum_{n \in \mathbf{Z}} \zeta_n(t) S(n,h)(x),$$
$$z_{xh}(x,t) = \sum_{n \in \mathbf{Z}} \tilde{\zeta}_n(t) S(n,h)(x),$$
$$b_h(x) = \sum_{n \in \mathbf{Z}} \beta_n S(n,h)(x),$$
$$u_{0h}(x,y) = \sum_{m,n \in \mathbf{Z}} \nu_{mn} S(m,h)(x) S(n,h)(y).$$

Using the preceding functions, we can construct, in a similar way as in Lemma 6.1, a function $F_h(.;z_h,z_{xh},b_h,u_{0h})$ such that

$$\|F - F_h\|_{L^2(\mathbf{R}^2)} \longrightarrow 0 \qquad \text{as } h \to 0. \qquad (7.76)$$

The details of calculations are omitted. By (7.76), we can, using the result of Lemma 7.1, construct a regularized solution of (7.69).

7.4 Notes and remarks

We have considered the backward heat problem in the special case of two space dimensions and under the condition that the support of the (unknown) initial temperature is contained in the quadrant $x \geq 0$, $y \geq 0$. Under these assumptions, we have been able to apply the results of Chapter 4 on the Hausdorff moment problem to derive explicit error estimates for the regularized solutions. For the case of a bounded domain Ω with zero Dirichlet condition on $\partial\Omega$, the problem has been regularized by various methods: truncated eigenvalue expansion, quasi-reversibility, Sobolev regularization, integral method coupled with Tikhonov regularization and others.

The problem of the backward heat equation on a bounded domain Ω with a regular boundary $\partial\Omega$ corresponding to unilateral conditions on the temperature function u, i.e.,

$$u \geq 0, \ \frac{\partial u}{\partial n} \geq 0, \ u\frac{\partial u}{\partial n} = 0 \text{ on } \partial\Omega$$

was raised in Payne [Pa]. As pointed out in [An2], a natural way to look at the problem would consist in converting it into a problem involving zero Neumann condition. In fact, as proposed in [An2], we let $v = u^2$. Then, v satisfies the equation

$$v_t - \Delta v = -\frac{|\nabla v|^2}{2v}$$

under the constraint $v \geq 0$, $u = \sqrt{v} \geq 0$, $\frac{\partial u}{\partial n} \geq 0$ and subject to the boundary condition

$$\frac{\partial v}{\partial n} = 0 \qquad \text{on } \partial\Omega$$

and terminal condition

$$v(x, 1) = g^2(1).$$

Thus the problem is converted to a unilateral problem for a semilinear parabolic equation with zero boundary condition.

We next consider another version of the "borehole" problem. Instead of the problem of surface temperature determination from the temperature measured at an interior point of the Earth, represented by a half-plane, as was done in the main text, we can consider the problem of determining the heat flux history through a space vehicle represented by a slab $0 \leq x \leq 1$, with the flux specified to be zero along the side $x = 1$, the flux through the side $x = 0$ being to be determined from the temperature measured at an interior point x_1, $0 < x_1 < 1$ at discrete times $t_0 < t_1 < ... < t_n$ (cf [BBS]). The problem can be formulated as a moment problem.

The two dimensional inverse Stefan problem was first studied by Colton [Co1], [Co2] under rather stringent regularity conditions. In [APT3], regularity conditions are relaxed and moreover, existence of an exact solution is not assumed. Moreover, in the latter work, the problem is regularized and error estimates are given for regularized solutions. In Colton and Reemtsen [CR] the problem is treated numerically.

8 Epilogue

Nonlinear moment problems: an example from gravimetry

In the preceding chapters, the moment problems considered are all linear, in fact, they are of the form

$$\int_{\Omega} v(x)d\sigma_n(x) = \mu_n, \qquad n = 1, 2, ..., \tag{8.1}$$

where Ω is a domain in $\mathbf{R}^k$, $d\sigma_n(x)$ is a given measure on Ω, $n = 1, 2, ...$, and $v(x)$ is a function on Ω to be determined. In this chapter, we shall consider a nonlinear moment problem arising in Gravimetry. The nonlinear problem consists of a sequence of equations

$$\int_{\Omega} K(x_n, v(x))dx = \mu_n, \qquad n = 1, 2, ..., \tag{8.2}$$

where K is nonlinear in the unknown function $v(x)$. Before giving an explicit expression for K, we deem it appropriate to explain the physical model.

The determination of the shape and location of an object Ω in the interior of the Earth, the density of which differs from that of the surrounding medium, is a fundamental problem of Applied Geophysics, in fact, belongs to Gravimetry, a branch of Geophysics concerned with the gravity fields in and around the Earth. Gravimetric methods are used for the identification of density inhomogeneities of the Earth. They consist in measuring the gravity anomalies or the gravity gradients created on the Earth's surface by the difference in density. The gravity gradient method presents some advantages over the gravity approach, as shown in [To].

Consider the Earth represented by the half-plane (x, y), $-\infty < y < H$ with $H > 0$ and let the body Ω be represented by $0 \le y \le \sigma(x)$, $0 \le x \le 1$. We assume that the unknown function $\sigma(x)$ is continuous and such that $\sigma(x) < H$ for $0 \le x \le 1$, $\sigma(0) = \sigma(1) = 0$.

Let the relative density of Ω, i.e., the difference between the density of Ω and that of the surrounding medium, be denoted by ρ, which we take to be a constant. Let $U = U(x, y)$ be the gravity potential created by ρ, i.e.,

$$U(x, y) = \frac{\rho}{2\pi} \int_{\Omega} \ln((x - \xi)^2 + (y - \eta)^2)^{-1/2} d\xi d\eta$$

$$= \frac{\rho}{2\pi} \int_0^1 \int_0^{\sigma(\xi)} \ln((x-\xi)^2 + (y-\eta)^2)^{-1/2} d\eta d\xi.$$

Then the vertical component of the gravity gradient created by ρ on the surface $y = H$ is

$$-\frac{\partial^2 U}{\partial y^2}\Big|_{y=H} = -\frac{\rho}{2\pi} \left\{ \int_0^1 \frac{(H-\sigma(\xi))d\xi}{(x-\xi)^2 + (H-\sigma(\xi))^2} - \int_0^1 \frac{H d\xi}{(x-\xi)^2 + H^2} \right\}. \quad (8.3)$$

Denoting by $f_0(x)$ the gravity gradient on the surface $y = H$, and taking $\rho = 1$, we have from (8.3), after some rearrangements,

$$\frac{1}{2\pi} \int_0^1 \frac{g(\xi)d\xi}{(x-\xi)^2 + g^2(\xi)} = f(x) \quad (8.4)$$

where we have set $g(\xi) = H - \sigma(\xi)$ and

$$f(x) = -f_0(x) - \frac{H}{2\pi} \int_0^1 \frac{d\xi}{(x-\xi)^2 + H^2}. \quad (8.5)$$

The equation (8.4) is a nonlinear integral equation of first kind for the determination of the unknown (continuous) function $g(x)$. The above formulation of the problem follows closely the presentation in [ANT], where it is shown that (8.4) admits at most one solution $g(x)$ continuous on $[0,1]$ such that $g(0) = g(1) = H$ and

$$0 < H - \alpha < g(x) < H \qquad \text{for } 0 < x < 1, \quad (8.6)$$

$\alpha > 0$ being a known constant. See also [AGV3].

It can be shown that the problem of finding $g(x)$ from (8.4) is ill-posed. It could be regularized by finite dimensional approximation, following the method of [AGV4]. We shall, instead, convert it into a nonlinear moment problem as follows. Since g is continuous and strictly positive on $[0,1]$, the function defined by the integral in the left hand side of (8.4) can be extended to a complex analytic function on a strip around the x-axis of the complex plane. Hence it is completely determined by its values at any bounded real sequence (x_n) with $x_i \neq x_j$ for $i \neq j$. Thus the integral equation (8.4) is equivalent to the (nonlinear) moment problem

$$\int_0^1 \frac{g(\xi)d\xi}{(x_n - \xi)^2 + g^2(\xi)} = 2\pi f(x_n), \qquad n = 1, 2, \dots . \quad (8.7)$$

We thus deal with a nonlinear mapping from a subset of the function space $L^2(0,1)$ into l^2. The regularization problem for such mappings is a wide open subject.

One of the common methods for dealing with nonlinear equations is the linearization method. It consists in approximating the original equation by an appropriate linear equation. The equation (8.4) offers a simple (and interesting) example of the linearization method.

As pointed out above, the function on the LHS of (8.4) can be extended to a complex analytic function on a strip of width $< H - \alpha$ around the real axis of the complex plane. Hence, it is completely determined by its values on an interval $(-\infty, -M)$ for any $M > 0$, i.e., Eq. (8.4) is equivalent to the equation

$$\int_0^1 \frac{g(\xi)d\xi}{(x-\xi)^2 + g^2(\xi)} = 2\pi f, \qquad x \le -M. \tag{8.8}$$

Now, for large $M > 0$ and $x \ge 0$, we have the following expansion of the LHS of (3.11)

$$\frac{g(\xi)}{(M+x+\xi)^2 + g^2(\xi)} = \frac{g(\xi)}{(M+x+\xi)^2 \left(1 + \frac{g^2(\xi)}{(M+x+\xi)^2}\right)}$$

$$= \frac{g(\xi)}{(M+x+\xi)^2} - \frac{g^3(\xi)}{(M+x+\xi)^4} + \dots$$

As a first approximation, we take

$$\frac{g(\xi)}{(M+x+\xi)^2 + g^2(\xi)} \approx \frac{g(\xi)}{(M+x+\xi)^2}$$

and consider the linear integral equation in g

$$\int_0^1 \frac{g(\xi)d\xi}{(M+x+\xi)^2} = 2\pi f(-M-x), \qquad x > 0. \tag{8.9}$$

By taking $x = 1, 2, \dots$, we get the equivalent moment problem

$$\int_0^1 \frac{g(\xi)d\xi}{(M+n+\xi)^2} = 2\pi f(-M-n) \equiv \mu_n, \qquad n = 1, 2, \dots. \tag{8.10}$$

As shown in [ANT], this moment problem admits at most one continuous solution $g(x)$ in $[0, 1]$.

References

[A]   G. Anger, *Inverse Problems in Differential Equations*, Akademie-Verlag, Berlin,1990.

[AGl]   N. I. Akhiezer and I. M. Glazman, *Theory of Linear Operators in Hilbert Space*, Vol II, Frederick Ungar Publishing Co., New York, 1961, pp 1-5.

[AGT]   D. D. Ang, R. Gorenflo and D. D. Trong, A multidimensional Hausdorff moment problem: regularization by finite moments, *Zeitschrift für Anal. und ihre Anwendungen* **18**, N^o1, 1999, pp 13-25.

[AGV1] D. D. Ang, R. Gorenflo and L. K. Vy, Backus-Gilbert regularization of a moment problem, Fachbereich Mathematik, Freie Univ. Berlin, *Preprint* A-7/93, 1993.

[AGV2] D. D. Ang, R. Gorenflo and L. K. Vy, Two regularization methods for the moment problem, Fachbereich Mathematik, Freie Univ. Berlin, *Preprint* A-19/93, 1993.

[AGV3] D. D. Ang, R. Gorenflo and L. K. Vy, A uniqueness theorem for a nonlinear integral equation of gravimetry, *Proceedings*, In: V. Lakshmikantham (ed.): World Congress of Nonlinear Analysts,'92, Vol III, pp. 2423-2430, W. de Gruyter, Berlin, 1996.

[AGV4] D. D. Ang, R. Gorenflo and L. K. Vy, Regularization of a nonlinear integral equation of gravimetry, *J. Inv. and Ill-posed Problems* **5** no2, 1997, pp. 101-116.

[AH]   D. D. Ang and D. D. Hai. On the backward heat equations, *Annales Polonici Mathematici* LII, 1990.

[Ak]   N. I. Akhiezer, *The Classical Moment Problem*, Hafner Publ. Co, 1965.

[Al]   G. Alessandrini, An extrapolation problem for harmonic functions, *Boll. UMI* **17B**, 1980, pp. 860-875.

[ALS]   D. D. Ang, J. L. Lund and F. Stenger, Complex variable and regularization methods of inversion of the Laplace transform, *Math. Comp.* **83**, 1989, pp. 589-608.

[An1]   D. D. Ang, Stabilized approximate solutions of the inverse time problem for a parabolic evolution equation, *J. Math. Anal. and Appl.*, Vol. 111, 1985, 148-155.

[An2]   D. D. Ang, On the backward parabolic equation: a critical survey of some current methods in numerical analysis and mathematical modelling, *Banach Center Publications*, **24**, Warsaw, 1990.

[ANT]   D. D. Ang, N. V. Nhan and D. N. Thanh, A nonlinear integral equation of gravimetry, uniqueness and approximation by linear moments, *Vietnam J. Math.*, Springer, **27**, n° 1, 1999, pp. 61-67.

[APT1] D. D. Ang, A. Pham Ngoc Dinh, D. N. Thanh, An inverse Stefan problem: identification of boundary value, *J. of Comp. and Appl. Math.* **66**, 1996, pp. 75-84.

[APT2] D. D. Ang, A. Pham Ngoc Dinh, D. N. Thanh, Regularization of an inverse Stefan problem, *J. of Diff. and Integ. Eqs.* **9** (1996), 371-380.

[APT3] D. D. Ang, A. Pham Ngoc Dinh, D. N. Thanh, A bidimensional inverse Stefan problem: identification of boundary value, *J. of Comp. and Appl. Math.*, **80**, 1997, 227-240.

[APT4] D. D. Ang, A. Pham Ngoc Dinh, D. N. Thanh, Regularization of a two-dimesional two phase inverse Stefan problem, *Inverse Problems* **13**, 1997, 607-619.

[AS]   M. Abramowitz and I.A. Stegun, *Handbook of Mathematical Functions*, Dover, New York, 1972.

[ASa]   R. S. Anderssen and V. A. Saull, Surface temperature history determination from borehole measurements, *Mathematical Geology* **5**, 1973, pp. 269-283.

[ASS]   R. Askey, I. J. Schoenberg and A. Sharma, Hausdorff's moment problem and expansion in Legendre polynomials. *J. Math. Anal. Appl.* **86**, 1982, pp. 237-245.

[ATT]   D. D. Ang, D. N. Thanh and V. V. Thanh, Regularized solutions of a Cauchy problem for Laplace equation in an irregular strip, *J. of Integ. Eqs. and Appl.*, **5**, No 4, 1993, pp. 429-441.

[AVG]   D. D. Ang, L. K. Vy and R. Gorenflo, A regularization method for the moment problem, in *Inverse Problems: Principles and Applications in Geophysics, Technology and Medicine*, Math. Research vol. 74, Akademie Verlag, Berlin, 1993, pp. 37-45.

[BG]   G. Backus and F. Gilbert, The resolving power of gross earth data, *Geophys. J. R. Astronom. Society*, **16**,1968, pp. 169-205.

[Bar]   V. Barbu, The inverse one phase Stefan problem, *Diff. and Integ. Eqs.* **3**, 1990, 209-218.

[Bau]   J. Baumeister, *Stable Solutions of Inverse Problems*, Vieweg, 1987.

[BBS]   J. V. Beck, B. Blackwell and C. R. St. Clair, *Inverse Heat Conduction, Ill-posed Problems*, Wiley, New York, 1985.

[BDP]   M. Bertero, C. De Mol and E.R. Pike, Linear inverse problems with discrete data. I: general formulation and singular system analysis, *Inverse Problems* **1**, 1985, pp. 301-330.

[Bl]   R.J. Blakeley, *Potential Theory in Gravity and Magnetic Applications*, Cambridge Univ. Press, 1995.

[Br]   H. Brezis, *Analyse Fonctionelle, Théorie et Applications*, Masson, 1983.

[Bui]   H. D. Bui, *Inverse Problems in the Mechanics of Materials: an Introduction*, CRC Press, Inc., 1994.

[But]   P. L. Butzer, A survey of the Whittaker-Shannon expansion theory and some of its extensions, *J. Math. Res. Exposition* **3**, 1983, pp. 185-212.

[Ci]   P. Ciarlet, *The Finite Element Method for Elliptic Problems*, North-Holland, 1978.

[CHWY] J. Cheng, Y. Hon, T. Wei, M. Yamamoto, Numerical computation of a Cauchy problem for Laplace's equation, ZAMM, **81**, 2001, 665-674.

[CJ]   H. S. Carslaw and J. C. Jaeger, *Conduction of Heat in Solids*, Oxford Univ. Press, 1950.

[Co1]   D. Colton, The inverse Stefan problem, *Ber. Gesellsch. Math. Datenverab.*, **77**, 1973, 29-41.

[Co2]   D. Colton, The inverse Stefan problem for the heat equation in two space variables, *Mathematika* **21**, 1974, 282-286.

[Co3]   D. Colton, *Partial Differential Equations*, Random House, 1988.

[CR]   D. Colton and R. Reemtsen, The numerical solution of the inverse Stefan problem in two space variables, *SIAM J. Appl. Math.* **44**, 1984, pp. 996-1013.

[CM]   J. R. Cannon, K. Miller, Some problems in numerical analytic continuation, *SIAM J. Num. Analysis* **2**, 1965, pp. 87-98.

[Di]   L. E. Dickson, *History of the Theory of Numbers*, Vol. II, Diophantine Analysis, Chelsea Publishing Company, New York, 1952.

[DNH]   Dinh Nho Hào, *Methods for Inverse Heat Conduction Problems*, Vol. 43 of *Methoden und Verfahren der mathematischen Physik* (edited by B. Brosowski and E. Martensen), Peter Lang, Franfurt am Main, 1998.

[DS]   N. Dunford and J. Schwartz, *Linear Operators. Part I: General Theory*, Interscience Publishers, Inc., New York, 1967.

[Ed]   Erdelyi et al., *Tables of Integral Transforms*, Vol. 1, Mc Graw-Hill, New York, 1954.

[EG]   H. W. Engl and C. W. Groetsch (Eds.), *Inverse and Ill-posed Problems*, Academic Press, Orlando, 1987.

[Ew]   R. E. Ewing, The approximation of certain parabolic equations backward in time by Sobolev equations, *SIAM J. Math. Anal.*, Vol.6, 1975, pp. 283-294.

[Fr]   A. Friedman, *Partial Differential Equations of Parabolic Type*, Englewood Cliff, N.J., 1964.

[GAT]   R. Gorenflo, D. D. Ang and D. N. Thanh, Regularization of a two-dimensional inverse Stefan problem, *Proceedings*, International Workshop on Inverse Problems, Hochiminh City, Jan. 17-19, 1995, pp. 45-54.

[GV]   R. Gorenflo and S. Vessella, *Abel Integral Equations*, Lecture Notes in Mathematics 1461, Springer Verlag, Berlin, 1991.

[GK]   K. Grysa and H. Kaminski, On a time stop choice in solving inverse heat conduction problems, *ZAMM*, **66**, 1968, pp. 368-370.

[Gr1]   C. W. Groetsch, *The Theory of Tikhonov Regularization for Fredholm Equations of the First Kind*, Pitman, London, 1984.

[Gr2]   C. W. Groetsch, *Inverse Problems in the Mathematical Sciences*, Vieweg, 1993.

[HNT]   N.V. Huy, N.V. Nhan and D. D. Trong, Reconstruction of analytic functions on the unit disc from a sequence of moments: regularization and error estimates, to appear in *Acta Mathematica Vietnamica*.

[Ho]   K. Hoffman, *Banach Spaces of Analytic Functions*, Prentice-Hall Inc., Englewood Cliffs, N. J. , 1962.

[HSo]   H. Haario and E. Somersalo, The Backus-Gilbert methods revisited: background, implementation and examples, *Numer. Funct. Anal and Optimiz.***9**, 1987, pp.917-943.

178   References

[HSt]   E. Hewitt and K. Stromberg, *Real and Abstract Analysis*, Springer-Verlag, New York, Inc., 1965.

[HSp]   K. H. Hoffmann and J. Sprekels, Real-time control in a free boundary problem connected with the continuous casting of steel, K.-H. Hoffmann and W. Krabs (editors): *Optimal Control of Partial Differential Equations ISNM*, **68**, 127-143, Birkhäuser-Verlag, Basel, 1984.

[H]   B. Hofmann, *Mathematik Inverser Probleme*, B.G. Teubner, Stuttgart and Leipzig, 1999.

[In]   G. Inglese, Recent results in the study of the moment problem. In: A. Vogel et al (eds): *Theory and Practice of Geophysical Data Inversion*, pp. 73-84, Vieweg Verlag, Braunschweig 1992.

[Is1]   V. Isakov, *Inverse Problems for Partial Differential Equations*, Applied Mathematical Sciences 127, Springer Verlag, New York, 1998.

[Is2]   V. Isakov, *Inverse Source Problems*, Mathematical Surveys and Monographs 34, American Mathematical Society, Providence, 1990.

[Jo1]   P. Jochum, The numerical solutions of the inverse Stefan problem, *Numer. Math.* **34**, 1980, pp. 411-429.

[Jo2]   P. Jochum, The inverse Stefan problem as a problem of nonlinear approximation theory, *J. Approx. Theory*, **30**, 1980, pp. 81-89.

[K]   A. Kirsch, *An Introduction to the Mathematical Theory of Inverse Problems*, Springer-Verlag, New York, 1996.

[KN]   K. Kurpisz and A. J. Nowak, *Inverse Thermal Problems*, Computational Mechanics Publications, Southampton, UK and Boston, USA, 1995.

[KS]   M. V. Klibanov and F. Santosa, A computational quasi-reversibility method for Cauchy problems for Laplace's equations, *SIAM J. Appl. Math.* **51**, 1991, pp. 1653-1675.

[KSB]   A. Kirsch, B. Schomburg and G. Berendt, The Backus-Gilbert method, *Inverse Problems* 4, 1988, 771-783.

[Kr]   W. Krabs, *On Moment Theory and Controllability of One-Dimensional Vibrating Systems and Heating Processes, Lecture Notes in Control and Information Sciences*, **173**, Springer-Verlag, 1992.

[La]   H. J. Landau, The classical moment problem: Hilbertian proof, *J. Funct. Anal.* **38**, 1980, pp. 255-272.

[Le1]   T. T. Le, Surface temperature determination from borehole measurements: a two-dimensional model, *Proceedings,* International Conference on Analysis and Mechanics of Continuous Media, 1995, pp. 222-225.

[Le2]   T. T. Le, A theorem of optimal recovery: applications to approximation theory, *Proceedings,* International Workshop on Inverse Problems HCMC, Jan. 17-19, 1995, pp. 106-116.

[Li]   J. L. Lions, *Quelques Méthodes de Résolution des Problèmes aux Limites Non Linéaires*, Dunod, 1969.

[LL1]   R. Lattès and J. L. Lions, *Méthode de Quasi-reversibilité et Applications.* Dunod, Paris, 1968.

[LL2]   R. Lattès and J. L. Lions, *The Method of Quasi-Reversibility Applications to Partial Differential Equations*, Elsevier, New York, 1969.

[LN1]   T. T. Le and M. Navarro, Surface temperature determination from bore-hole measurements: regularization and error estimates, *Inter. J. Math.& Math. Sci.* **18**, 1995, 601-606.

[LN2]   T. T. Le and M. Navarro, More on surface temperature determination, *SEA Bull. Math.* **19**, $N°3$, 1995, 81-86.

[LNTT]   T. T. Le, N. V. Nhan, D. N. Thanh and D. D. Trong, A moment approach to some problems in heat conduction. In: R. Gorenflo and M.P. Navarro (eds): *Proceedings International Conference on Inverse Problems*, Manila, 1998, pp. 148-153, Matimyás Matematika, Special Issue, August 1998.

[LTT]   T. T. Le, D. N. Thanh and P. H. Tri, Surface temperature from borehole measurements: a finite slab model, *Acta Mathematica Vietnamica,* **20**, 1995, pp. 193-206.

[LS]   L. A. Liusternik and V. J. Sobolev, *Elements of Functional Analysis*, Frederick Ungar Publishing Co., New York, 1961.

[Mi]   V.P. Mikhailov, *Partial differential equations*, Mir, Moscow, 1978.

[ML]   G. G. Margaril-Il'yaev and T. T. Le, On the problem of optimal recovery of functionals, *Russian Math. Surveys* **42**, $n°2$, March-April, 1987.

[MSM]   G. D. Mostow, J. H. Sampson, J. P. Meyer, *Fundamental Structures of Algebra*, Mc Graw-Hill, 1963.

[Na]   F. Natterer, Numerical treatment of ill-posed problems, in [Tal1], pp. 142-167.

[NC]   N. V. Nhan and N. Cam, The backward heat equation: regularization by cardinal series, *Scientific Bull.*, College of Education, Hochiminh City, 1999.

[Pa]   L. E. Payne, Improperly posed problems in partial diiferential equations, *Regional Conf. SIAM*, 1975.

[Pe]   R. N. Pederson, On the unique continuation theorem for certain second and fourth order elliptic equations, *Comm. on Pure and Appl. Math.*, Vol. XI, 67-80, 1958.

[Re]   H. J. Reinhardt, Regularization and approximation of Cauchy problem for elliptic equations, *Abstracts*, International Conference on Inverse Problems and Applications, Feb. 23-27,1998, Manila, Philippines.

[RN]   F. Riesz and B. Sz-Nagy, *Functional Analysis*, Frederick Ungar Publishing Co., New York, 1955, pp 115-118.

[Ru]   W. Rudin, *Real and Complex Analysis*, McGraw-Hill, 1987.

[Sa]   G.Sansone, *Orthogonal Functions*, Interscience, N.Y., 1959.

[Sha]   C. E. Shannon, A mathematical theory of communication, *Bell System Tel. Tech. J.* **27**, 1948, pp 379-423 and 623-656.

[Sho1]   R. Showalter, Quasi-reversibility of first and second order parabolic evolution equations, in A.Carasso and A. P. Stone (eds.) *Improperly Posed Boundary Value Problems*, pp. 76-84, Pitman Research Notes in Math. Vol. 1, London, San Francisco, Melbourn, 1975.

[Sho2]   R. Showalter, The final value problem for evolution equation, *J. of Math. Anal. and Appl.* **47**, 1974, pp. 563-572.

[St]   F. Stenger, *Numerical Methods Based on Sinc and Analytic Functions*, Springer-Verlag, 1993.

180   References

[STa]   J. A. Shohat and J. D. Tamarkin, *The Problem of Moments*, AMS Math. *Surveys*, Providence R I, 1943.

[STi]   R. Showalter and T. W. Ting. Pseudo-parabolic partial differential equations, *SIAM J. Math. Anal.* **1**, 1970, 1-26.

[Tal1]   G. Talenti, *Inverse problems*, LNM 1225, Springer-Verlag, New York, 1986.

[Tal2]   G. Talenti, Recovering a function from a finite number of moments, *Inverse Problems* **3**, 1987, pp. 501-517.

[Tay]   A. E. Taylor, *Advanced Calculus*, Blaisdell Publishing Company, 1965.

[Tay1]   J. M. Taylor, The condition number of Gram matrix and related problems, *Proc. of the Royal Soc. of Edinburg*, **80A**, 1978, pp.45-56.

[Th]   V. Thomée, *Galerkin Finite Element Method for Parabolic Problems*, Lect. Notes in Math., Springer, 1984.

[TAn]   D. D. Trong and D. D. Ang, Reconstruction of analytic functions: regularization and optimal recovery, *Preprint*, 1997.

[TAr]   A. N. Tikhonov and V. Y. Arsenin, *Solutions of Ill-posed Problems*, Winston and Sons, Washington, 1977.

[To]   W. Torge, *Gravimetry*, W. de Gruyter, Berlin, 1989.

[Wa]   P. K. C. Wang, Control of a distributed parameter system with a free boundary, *Int. J. Control*, Vol. 5,N°4, 1967, 317-32

[Wh]   E. T. Whittaker, On the functions which are represented by expansion of the interpolation theory, *Proc. Roy. Soc. Edinburgh* **35**, 1915, pp 181-194.

[Vo]   Y. V. Vorobiev, *Method of Moment in Applied Mathematics*, Gordon and Breach Science Publishers (1965).

Index

Druck: Strauss Offsetdruck, Mörlenbach
Verarbeitung: Schäffer, Grünstadt

4. Lecture Notes are printed by photo-offset from the master-copy delivered in camera-ready form by the authors. Springer-Verlag provides technical instructions for the preparation of manuscripts. Macro packages in T_EX, L^AT_EX2e, $L^AT_EX2.09$ are available from Springer's web-pages at

http://www.springer.de/math/authors/b-tex.html.

Careful preparation of the manuscripts will help keep production time short and ensure satisfactory appearance of the finished book.

The actual production of a Lecture Notes volume takes approximately 12 weeks.

5. Authors receive a total of 50 free copies of their volume, but no royalties. They are entitled to a discount of 33.3 % on the price of Springer books purchase for their personal use, if ordering directly from Springer-Verlag.

Commitment to publish is made by letter of intent rather than by signing a formal contract. Springer-Verlag secures the copyright for each volume. Authors are free to reuse material contained in their LNM volumes in later publications: A brief written (or e-mail) request for formal permission is sufficient.

Addresses:

Professor Jean-Michel Morel
CMLA, École Normale Supérieure de Cachan
61 Avenue du Président Wilson
94235 Cachan Cedex France
e-mail: Jean-Michel.Morel@cmla.ens-cachan.fr

Professor Bernard Teissier
Institut de Mathématiques de Jussieu
Equipe "Géométrie et Dynamique"
175 rue du Chevaleret
75013 PARIS
e-mail: Teissier@ens.fr

Professor F. Takens, Mathematisch Instituut
Rijksuniversiteit Groningen, Postbus 800
9700 AV Groningen, The Netherlands
e-mail: F.Takens@math.rug.nl

Springer-Verlag, Mathematics Editorial, Tiergartenstr. 17
D-69121 Heidelberg, Germany
Tel.: +49 (6221) 487-701
Fax: +49 (6221) 487-355
e-mail: lnm@Springer.de